Advances in Intelligent Systems and Computing

Volume 421

Series editor

Janusz Kacprzyk, Polish Academy of Sciences, Warsaw, Poland
e-mail: kacprzyk@ibspan.waw.pl

About this Series

The series "Advances in Intelligent Systems and Computing" contains publications on theory, applications, and design methods of Intelligent Systems and Intelligent Computing. Virtually all disciplines such as engineering, natural sciences, computer and information science, ICT, economics, business, e-commerce, environment, healthcare, life science are covered. The list of topics spans all the areas of modern intelligent systems and computing.

The publications within "Advances in Intelligent Systems and Computing" are primarily textbooks and proceedings of important conferences, symposia and congresses. They cover significant recent developments in the field, both of a foundational and applicable character. An important characteristic feature of the series is the short publication time and world-wide distribution. This permits a rapid and broad dissemination of research results.

More information about this series at http://www.springer.com/series/11156

Roman Szewczyk · Ivan Kaštelan
Miodrag Temerinac · Moshe Barak
Vlado Sruk
Editors

Embedded Engineering Education

Editors
Roman Szewczyk
Industrial Research Institute for Automation and Measurements PIAP
Warsaw
Poland

Ivan Kaštelan
Department of Computing and Control Engineering
University of Novi Sad
Novi Sad
Serbia

Miodrag Temerinac
Department of Computing and Control Engineering
University of Novi Sad
Novi Sad
Serbia

Moshe Barak
Department of Science and Technology Education
Ben-Gurion University of the Negev
Beersheba
Israel

Vlado Sruk
Faculty of Electrical Engineering and Computing
University of Zagreb
Zagreb
Croatia

ISSN 2194-5357 ISSN 2194-5365 (electronic)
Advances in Intelligent Systems and Computing
ISBN 978-3-319-27539-0 ISBN 978-3-319-27540-6 (eBook)
DOI 10.1007/978-3-319-27540-6

Library of Congress Control Number: 2015958858

Printed on acid-free paper

This Springer imprint is published by SpringerNature
The registered company is Springer International Publishing AG Switzerland

Preface

Rising development of computer-based applications in almost all fields, from medicine up to spacecraft technology, opened needs for new specialists—embedded engineers. Embedded engineering covers different application fields solving needed functionality by embedded computer technology. Therefore, the main challenges in embedded engineering education are interdisciplinary approach and fast development of computer technology with variety of customized processor platforms.

Some aspects of embedded engineering education have been the subject of the European research project "FP7-ICT-2011-8/317882: Embedded Engineering Learning Platform." The five achieved outcomes—unified platform, basic set of exercises, augmented reality interface, remote laboratory, and evaluation methodology with tools—can be a valuable contribution for the establishment of an embedded engineering profile.

This book is initiated by few universities from different countries which try to establish an attractive and efficient study program for embedded engineering. Besides results and experiences from the research project, it also includes some other relevant experiences and expertise.

The book contains 12 original contributions.

The first one is discussing potential answers to the main challenges of embedded engineering education. Some appropriate approaches for the definition of study programs and their evaluation are given, together with early experiences at the university where this approach is applied in last five years.

The second chapter presents a unified learning platform for embedded engineering which includes hardware and software modules necessary for laboratory exercises during the whole curriculum. The main goal is significant reduction of introduction overheads in different courses. The platform has been evaluated at four universities.

The next chapter gives a comprehensive overview of the basic set of 65 exercises which accompany the unified learning platform. Besides documentation for each exercise, necessary support and check routines are included.

The fourth chapter gives an educational approach regarding advanced computer architectures. Implemented Cray-1 architecture on the unified learning platform allows students to learn, through exercises, the historical development of multicore systems and supercomputers.

The fifth chapter presents new improvement concepts in education using augmented reality technology. Use scenarios, field studies, activity analysis, and formative evaluations are presented.

The sixth chapter presents development and usage advances of an augmented reality interface for the unified learning platform. This interface is designed as the learning assistance module providing students an easy and broad access to additional information during the laboratory exercises.

The seventh chapter proposes a very useful remote laboratory concept based on the usage of the unified platform. Necessary hardware and software modules are described, and different usage scenarios are evaluated.

The eighth chapter gives examples of advanced projects on the unified learning platform evaluating their impact on embedded engineering education. Besides standard laboratory exercises, the advanced student projects provide the best way to familiarize students with interdisciplinary embedded applications.

The ninth chapter gives a useful overview of experiences at the university where the remote laboratory has been used in running study programs. Additionally, the given evaluation provides very important guidelines for further usage of the remote laboratory on the unified platform.

The tenth chapter proposes adequate evaluation methods and tools for embedded engineering education. Particularly, exploring aspects of self-regulated learning is important for higher study efficiency.

The eleventh chapter addresses a very important aspect of student motivation. The proposed approach is based on specially designed introductory projects in electrical and computer engineering. This approach is illustrated through an example—team project for planning a window-cleaning robot.

The last chapter is focused on interfacing in intelligent sensor networks. Analyzing different solutions, a new concept has been proposed where recognizing task intelligence is suited on sensor.

The editors hope that this book will open a broader discussion about necessary knowledge and appropriate learning methods for the new profile of embedded engineers.

Roman Szewczyk
Ivan Kaštelan
Miodrag Temerinac
Moshe Barak
Vlado Sruk

E2LP Digest

Embedded systems have been furiously developed in last years. Therefore, the embedded engineering education play important role in all universities. The main challenges are interdisciplinary approach and high development dynamics.

This book focuses on the outcome of the E2LP project presented through the scientific–technical contributions and aims to be available to a wide scientific community. Additionally, some experiences and researches outside this project have been included. This book provides information about the achieved results of the E2LP project as well as some broader views about the embedded engineering education. It captures project results and applications, methodologies, and evaluations, leads us to the history of computer architectures, brings us a touch of the future in education tools, and provides a valuable resource for anyone interested in embedded engineering education concepts, experiences and material.

E2LP project deals with challenges in engineering education for embedded systems at university level which requires a complex and multidisciplinary approach involving the understanding of various systems based on different technologies and system solution optimizations. The main idea behind the project is to provide a unified platform which will cover a complete process for embedded systems learning. A modular approach is considered for skills practiced through supporting individualization in learning. This platform shall facilitate a novel development of universal approach in creative learning environment and knowledge management that encourages the use of ICT. New learning model is challenging the education of engineers in embedded systems design through real-time experiments that stimulate curiosity with ultimate goal to support students to understand and construct their personal conceptual knowledge based on experiments. In addition to the technological approach, the use of cognitive theories on how people learn will help students to achieve a stronger and smarter adaptation of the subject. Applied methodology requires evaluation from the scientific point of view in parallel with the implementation in order to feedback results to the R&D.

As a result, the proposed Embedded Computer Engineering Learning Platform will help to educate a sufficient number of future engineers in Europe, capable of designing complex systems and maintaining a leadership in the area of embedded systems, thereby ensuring that our strongholds in automotive, avionics, industrial automation, mobile communications, telecoms, and medical systems are able to develop. In such a manner, the E2LP intends to increase European competitiveness in the learning process of embedded computer engineering, ensuring further technological and methodological development of the educational approach in this field. In-depth information about the project can be found on the project Web page http://www.e2lp.org.

Contents

Challenges in Embedded Engineering Education 1
Ivan Kastelan, Nikola Teslic and Miodrag Temerinac

Unified Learning Platform for Embedded Engineering 29
Ivan Kastelan, Nikola Teslic and Miodrag Temerinac

Exercises for Embedded Engineering Learning Platform 45
Branka Medved Rogina, Karolj Skala, Peter Škoda, Ivan Sović and Ivan Michieli

Implementation of Advanced Historical Computer Architectures. 61
Zorislav Šojat, Karolj Skala, Branka Medved Rogina, Peter Škoda and Ivan Sović

Methods for User Involvement in the Design of Augmented Reality Systems for Engineering Education. 81
Margarita Anastassova, Sabrina Panëels and Florent Souvestre

Augmented Reality Interface for E2LP: Assistance in Electronic Laboratories Through Augmented Reality. 93
Enara Artetxe González, Florent Souvestre and Jorge R. López Benito

E2LP Remote Laboratory: e-Learning Service for Embedded Systems Education 109
Rafał Kłoda and Jan Piwiński

Advanced Projects and Applications for Embedded Systems Engineering on E2LP Platform 119
Dario Grgić, Sebastian Böttcher, Marc Pfeifer, Johannes Scherle, Benjamin Völker, Jan Burchard, Sebastian Sester and Leonhard M. Reindl

E2LP Remote Laboratory: Introduction Course and Evaluation at Warsaw University of Technology. 133
Rafał Kłoda, Jan Piwiński and Roman Szewczyk

Exploring Aspects of Self-regulated Learning Among Engineering Students Learning Digital System Design in the FPGA Environment—Methodology and Findings 139
Moshe Barak, Ivan Kastelan and Zvi Azia

Is It Possible to Increase Motivation for Study Among Sophomore Electrical and Computer Engineering Students? 161
Aharon Gero

Interrupts Become Features: Using On-Sensor Intelligence for Recognition Tasks 171
Kristof Van Laerhoven and Philipp M. Scholl

About the Editors

Professor Roman Szewczyk received both his Ph.D. and D.Sc. in the field of mechatronics. He is specializing in the modelling of properties of magnetic materials as well as in sensors and sensor interfacing, in particular magnetic sensors for security applications. He is the leading the development of a sensing unit for a mobile robot developed for the Polish Police Central Forensic Laboratory and of methods of non-destructive testing based on the magnetoelastic effect. Professor Szewczyk was involved in over 10 European Union-funded research projects within the FP6 and FP7 as well as projects financed by the European Defence Organization. Moreover, he was leading two regional- and national-scale technological foresight projects and was active in the organization and implementation of technological transfer between companies and research institutes.

Roman Szewczyk is Secretary for Scientific Affairs in the Industrial Research Institute for Automation and Measurements (PIAP). He is also Associate Professor at the Faculty of Mechatronics, Warsaw University of Technology and a Vice-chairman of the Academy of Young Researchers of the Polish Academy of Sciences.

Ivan Kaštelan born in 1985 and received the B.Sc., M.Sc., and Ph.D. degrees in electrical and computer engineering from the Faculty of Technical Sciences, University of Novi Sad, Serbia, in 2008, 2009, and 2014, respectively. He received the award for the best student of the Faculty of Technical Sciences in 2008 and the award for his teaching work in 2014. He is currently an Assistant Professor at the Department of Computing and System Control. He is also working as a software engineer in RT-RK Institute for Computer Based Systems. During his M.Sc. studies, he was an intern in Micronas GmbH where he worked in the field of synthesizable assertions for verification of digital systems. He was an exchange Ph.D. student in Universität Freiburg for 4 months where he worked in the field of channel modelling. His current research interests include digital and computer system design, signal processing, and algorithms. During his Ph.D. studies, he worked on the hardware and algorithms for automated verification of digital television sets and touch screen-based devices. He is a coordinator deputy and work package leader in the

EU FP7 project "E2LP—Embedded Engineering Learning Platform." He published around 50 papers in international journals and conferences. He is also a coauthor of 2 patents.

Miodrag Temerinac received the Ph.D. degree in electrical and computer engineering from the University of Belgrade, Serbia, in 1983.

From 1976 to 1992, he was with the Faculty of Technical Sciences of the University of Novi Sad, Serbia, as the full professor heading the chair of communications and as the vice faculty dean for research. He is Alexander-von Humboldt fellow (1988–1990) doing research in fields of audio and video compression at the University of Hannover in Germany. In 1992, he moved to industry joining the semiconductor company Micronas GmbH in Freiburg, Germany, where he worked on the IC development for consumer electronics and later as the manager for R&D external relations. In 2005–2006, he founded Micronas R&D Center in Shanghai spending two years as the director of system development. Also, he founded and headed the Micronas development center for TV software in Novi Sad from 2007 to 2009. In his last years, he is again with the University of Novi Sad heading the group for computer engineering and communication. He was the cofounder of the RT-RK company.

His fields of interest are DSP algorithms and architectures, audio and video signal processing, video quality assessment, hw/sw codesign of complex systems on chip, product development in consumer electronics, knowledge management, and management of internal and external development networks. He is the senior IEEE member and the VDE/ITG member.

Prof. Moshe Barak is a Professor at the Department of Science and Technology Education, Ben-Gurion University of the Negev, Israel. His background is in electrical and electronics engineering, and he received his Ph.D. degree in science and technology education from the Technion—Israel Institute of Technology (1986). Barak's research interests focus on fostering higher-order cognitive skills such as problem solving and creativity in technology and engineering education, teaching advanced interdisciplinary technological subjects such as control systems and robotics to children, and using information and communication technology (ICT) in education and evaluation of educational programs.

Vlado Sruk is an Associate Professor at the Department of Electronics, Microelectronics, Computer and Intelligent Systems at Faculty of Electrical Engineering and Computing. He obtained Ph.D. degree in computer science from the University of Zagreb in 1998. He participated as a researcher in international scientific projects. His current research interests are in the areas of multicore embedded systems, mobile computing, high-performance computing with emphasis on memory architectures, and state-of-the-art concepts and techniques in multicore software engineering and fault-tolerant computing.

He is a member of IEEE and ACM society. He participates in conference international programs committees, and he serves as a technical reviewer for various international journals and conferences.

Challenges in Embedded Engineering Education

Ivan Kastelan, Nikola Teslic and Miodrag Temerinac

Abstract Tremendous progress in semiconductor technology enabled broad proliferation of computer-based solutions in almost all fields, from medical instrumentation up to spacecraft technology. This development opened the question about a new engineering profile being able to act in research and development of embedded systems. Critical parts of this curriculum are the learning platform, embedded software education, methodologies and tools in education, evaluation procedure, recognition of profile potentials and dropout rate. In this paper a contribution to the structuring of embedded engineering education is given. Three education approaches relevant for this proposal are analyzed and the structured approach of the whole education process is described. Case study, where this concept is partially applied, concludes the paper.

Keywords Embedded engineering · Education · Curriculum

1 Introduction

Tremendous progress in semiconductor technology enabled broad proliferation of computer-based solutions in almost all fields, from medical instrumentation up to spacecraft technology. This approach gained special visibility of mass users through broad consumer applications like smartphones, smart TV, smart home, driving assistants in car and finally Internet of Things (IoT) concepts. Common for

I. Kastelan (✉) · M. Temerinac
Faculty of Technical Sciences, University of Novi Sad,
Trg Dositeja Obradovica 6, 21000 Novi Sad, Serbia
e-mail: ivan.kastelan@rt-rk.uns.ac.rs

N. Teslic
RT-RK Institute for Computer Based Systems, Narodnog Fronta 23a,
21000 Novi Sad, Serbia

R. Szewczyk et al. (eds.), *Embedded Engineering Education*,
Advances in Intelligent Systems and Computing 421,
DOI 10.1007/978-3-319-27540-6_1

all these solutions is using processor modules with adequate software to solve user needs. This approach is broadly known as "embedded systems" (perhaps a better name would be "embedded intelligence").

This development opened the question about a new engineering profile being able to act in research and development of embedded systems [1]. In distinction to the classical "general purpose computer engineering education programs" the embedded software and system engineering should cover specific requirements of interdisciplinary usage. Generally, we are facing an intersection in knowledge between electrical engineering, computer engineering and computer science on the one side and specialist profiles in application on the other side fields (mechanical engineering, civil engineering, biology and medicine, etc.). The first dilemma is where to settle necessary embedded-based knowledge—either in engineering programs covering fields of usage or in the standard computer engineering education. The second approach is broadly driven in many initiatives [2–6]. The first approach is still not structurally evaluated. There are many heuristic approaches in many usage fields, simple usage of available computer-based tools, but not trough an evaluation which should be changed in corresponding education programs regarding possibilities of a computer-based approach. However, this is a classical "interfacing problem of communication", known in many technical systems but not so much explored in the educational context. The first step would be to establish an "interface terminology" as it is mentioned in [7].

System integration of many hardware and software modules in embedded systems makes testing and verification the key success factors in embedded solutions [8]. Therefore, testing and verification methodology reflects as a big challenge for embedded engineering education.

Focusing the necessary changes in computer engineering and computer science education, we are meeting the next dilemma—what are generic principles which should be covered in the embedded engineering education for variety of using fields. An intuitive approach is to ask industry about needed engineering profiles where educated engineers will act for their lifetime. A joint work between academy and industry certainly improves engineering education programs as it is mentioned in [9, 10]. But, this approach is facing an issue—the imbalance between technology cycle times and working life times. When technology cycle times, like in the past, are longer than the people working times, industry needs could perfectly define requirements for education which takes more than 10 years (about 12 years for basic and high school education plus about 5 years engineering education). Today, technology cycles are even shorter than the engineering education time, which means that the industry needs today could only express educational needs in the past [11]. Therefore, an interactive approach between industry and academy with improving loops inside education times will be needed. A concurrent evaluation of systems and processes mentioned in [12] could be a right answer to this challenge. Even an early impact during basic and high school education could help to overcome this race between education and development [13–15].

Talking about embedded engineering studies an important question is the learning platform. In practice embedded solutions are implemented on different platforms (hardware and software). The variety and dynamics of used platforms cannot be covered by a unified study program. Therefore, some kind of abstraction and generalization in engineering education has to be applied. A possible overcoming approach is proposed in [16] trying to leverage orthogonality, theoretical background and necessary education platforms. Having in mind a crucial point of the embedded systems approach, all problems should be solved by software approach, this platform broadly uses the unified API (application programming interface) approach. In [17] a unified learning platform for all embedded engineering related courses during the computer engineering studies is proposed using modular and extendable approach.

Especially critical part of embedded engineering education is related to the embedded software [18]. Having in mind limitations and requirements for embedded systems we are facing the problem of necessary knowledge which provides solutions on propitiatory platforms in different applications. Therefore, a unified operating system for embedded software doesn't exist [19]. For example, the embedded Linux has variety of versions in real life depending on a used platform. Also, the software architecture strongly depends on applications and available resources. Even a same functionality could be fully differently implemented on different platforms. For example, appearance of multi-core systems is reflected in education through specific courses of multi-core architecture design and parallel programming [20]. A proposal for software engineering education, given in [21], also points out the importance of processes, like coaching and team routines, towards planning and improving of capstone courses.

The next important question is which methodologies and tools should be involved in education of embedded engineers. In practice many proprietary platforms having own tools are used. In education different approaches should be summarized in a set of generic knowledge providing engineers successful acting with variety of used platforms [11]. This set should minimally include methodology and tools for real-time operating systems [22], interfacing concepts, networking protocols and application development. A promising approach is to teach generic principles but to train further engineers in practical exercises using actual tools [23]. Today, two most commonly used embedded environments for these purposes are embedded Linux [24] and Android platform for applications [15]. Beside the classical study approach (classroom and lab) the e-learning and remote lab approaches proposed in [25–28] could significantly improve engineering education. Also, using of augmented reality interfaces in lab exercises [29] contribute to an improvement of education process. An interesting approach regarding IoT programming is proposed in [14]. An important dilemma in education methodology is the basic concept—either standard approach starting with basic knowledge modules and building up targeted engineering profile or problem-oriented and project-based education [30]. Some approaches are fully application centered. For example the REACT tools for AI planning in robotics [31] provide an abstract platform based on a specific high-level programming language.

Evaluation and assessment of engineering education is an important point closing the improvement loop of education process. Here, a dilemma is what and how to measure—achieved knowledge level, usability of acquired knowledge, soft capabilities (team work, project management) as well as capability for innovations. Some approaches for education quality evaluation and assessment are proposed in [23, 32, 33]. The proposal [13] even covers a comprehensive education concept for embedded systems from the school up to university.

Finally, a big problem in embedded engineering education is low recognition of profile potentials by freshman decisions and high dropout rate during studies. Therefore, some initiatives in a better visibility of the embedded engineering profile as well as in a better motivation during the studies could be very helpful to increase efficiency of studies. Some proposals with experiences are published in [34] using gamification of learning units and [35] leveraging targeted skills and student motivation.

In this paper a contribution to the structuring of embedded engineering education is given. The second chapter briefly analyses three education approaches relevant for this proposal. Our approach starts in the third chapter with challenges which should be covered in embedded engineering education. The structured approach of the whole education process is described in the fourth chapter matching skills, knowledge and courses. Evaluation methodology is also included providing a closed improving loop in the education process. A case study, where this concept is partially applied, is given in fifth chapter. Conclusions and references are given in the last two chapters.

2 Overview of Relevant Education Programs

The chosen three education programs were the starting point for the proposed structural approach. The first one is perhaps the most comprehensive program for computer engineering and computer science education referenced in many actual study programs. The second one is an initiative of leading universities to develop a new model of engineering education. The third one deals with problem-based and project-based approach which improves efficiency and attractiveness of engineering studies.

The Computer Science Curricula [5] is the result of the ACM/IEEE Joint Task Force which included experts from more than 3000 relevant departments worldwide (about 1500 are in USA). The earlier report from the same ACM/IEEE joint task force covers the Computer Engineering Curricula [6]. The main goal was to provide curriculum guidelines for undergraduate studies in computer science. But, the used methodology and obtained results are also applicable for other engineering fields as well as for higher study levels. The published results include principles, knowledge areas, learning outcomes and course structure with ranking. Also, many exemplar courses and curricula are presented to promote easier sharing of educational ideas.

The applied principles are deviated from four high-level themes: big-tent view (open for cross-disciplinary work), managing the size of curriculum (rapid development covered rather with ongoing re-evaluation of essential topics than continuously expanding the curriculum size), actual course examples (identification and sharing of successful courses from existing curricula) and institutional needs (curriculum structure adaptations according to targeted outcomes and resource constraints).

The knowledge areas cover 18 disciplines for each curriculum. A significant part of them make overlapped disciplines, common for computer science and engineering. Differences are caused by different accents. The computer sciences are more focused to computation and its applications but the computer engineering is more oriented to the implementation having close items with electrical engineering. Having in mind interdisciplinary aspects of computer science and engineering the list of identified knowledge areas could be extended with some application-centric disciplines like mathematics, electrical and mechanical engineering, statistics, life sciences, psychology and fine arts. These extensions should develop the flexibility to work across disciplines.

Learning outcomes are defined as the capturing of various skills associated with obtained knowledge. The concept of learning outcomes considers not just achieved knowledge with relevant practical skills, bat also personal and transferable skills. A learning outcome can be associated with one or more knowledge areas.

Course structure is the final output for a study curriculum. Each course is associated with one or more knowledge areas and includes lectures and lab exercises which are necessary to achieve requested learning outcomes. A learning outcome can be considered in the design of more courses. Design of courses also defines ranking (study credit measures like ECTS in Europe or College Credits in USA) and time requirements (hours per lecture and exercise). Content of courses is generally defined by required learning outcomes but the kind of implementation may vary depending on geographical, cultural and institutional conditions. In the analyzed recommendations of engineering education all courses are grouped in core courses and elective courses. The core courses are subdivided into "Tier-1" (obligatory courses) and "Tier-2" (recommended courses). The group of elective courses depends on application and university specific requirements.

The considered two curriculum guidelines for computer science and computer engineering comprehensively describe methodology and curriculum structure broadly covering crucial knowledge areas. Two small drawbacks could be pointed out: insufficient consideration of soft engineering skills (team work, project management) and missed incorporation of innovative capabilities.

The CDIO [2, 36] is a common initiative of several universities in USA and Sweden to develop a new model of engineering education including evaluation at involved universities. The main goal is development of a comprehensive engineering education including all aspects of modern engineering: a Conceive-Design-Implement-Operate (CDIO) approach. This initiative covers a reform of engineering study curricula, development of corresponding pedagogy with necessary assessment scheme and establishment of workshop laboratories where multi-disciplinary student teams work on complex projects.

New proposed approach in engineering curricula include: five professional tracks (Researcher, System Engineer, Designer/Developer, Support Engineer and Engineering Manager), combination of lesson-based and block-based courses and project-based lab exercises.

The proposed reform of pedagogy reform at universities mainly deals with question how to learn in parallel with question what to learn. The proposed approach includes four efforts in engineering teaching: increase the scale of hands-on learning, move toward a style emphasizes problem formulation, increase the level of proactive engagement of students and provide functional feedback mechanisms.

Such reform of engineering education requires rigorous and comprehensive assessment providing systematical quality monitoring and improvement of education programs. The proposed approach covers four assessment sets: clear and measurable disciplinary goals (used Bloom's taxonomy), skills assessment (qualitative and quantitative measures), creativity assessment (portfolios, design reviews and desk critiques) and programmatic assessment (costs including metrics of student time, faculty time and infrastructure).

A special novelty in university practice is the introduction of workshop laboratories. The workshops should provide students authentic personal experience manipulating the tangible objects which relate to the theoretical lessons. Of course, the workshops should also promote teamwork and project management. Laboratories for workshops should be equipped for all phases of development: concept, design, implementation and operation. This approach should develop student skills regarding necessary knowledge, its application to solve given problems and testing of obtained solutions. Three types of facilities/activities have been considered: browsing laboratories, sharing-of-research laboratories and system/product laboratories.

The POPBL (problem-oriented and project-based learning) [30] introduces new learning approaches targeting development of innovative and creative thinking by students. In distinction to the traditional engineering syllabus, where subject-oriented knowledge modules define design of courses, the POPBL approach puts a problem in the focus of building up necessary knowledge and skills during the solving process. Application of this approach in study programs at Aalborg University (Denmark) showed significant improvements in the learning process. Education of engineers being able to handle sustainability-related problems requires ability for interplay, mix and diversity—all aspects considered in the POPBL approach. A drawback of such approach could be the necessary structural aspect of engineering education. Therefore, a balance between subject-oriented and problem-oriented approaches should be found by designing study programs.

Also, many larger companies provide a lot of contributions for engineering education, e.g. [37–39]. They provide excellent learning platforms (evaluation boards, SDK for software development) accompanied with on-line courses but always related only to their own product portfolio.

3 Challenges of Embedded Engineering Education

A generic model of embedded solutions would be very helpful to structure the embedded engineering education. Variety of embedded applications makes this abstraction very demanding. Our approach, shown in Fig. 1, could be a useful contribution in this direction.

Application fields cover almost all aspects of human activities (e.g. social, politics, economics, bio-medicine, health, energy, automotive, consumer etc.). An asked functionality can be expressed explicitly (like in automotive or consumer applications) or descriptively (like in social or health applications). For a later computer-based modeling the descriptive requirements are especially difficult to handle.

Interfacing between application fields and embedded solutions includes input and output data which must be precisely defined. The input data acquisition includes two processes: measurement (sensors) and acquisition (could be distributed what requires an acquisition network). The output data control can be used either for monitoring of processes/functionalities (displays) or as a feedback in an automatic process control (actors). Beside technical aspects the definition of this interfacing is a critical understanding challenge between experts in an application field and embedded engineers.

An embedded solution makes in the first step the model of a given problem including required functionality and available input/output interfaces. Additionally, the frame conditions (costs, power consumption, security etc.) are considered by the design of an appropriate hardware platform. This platform,

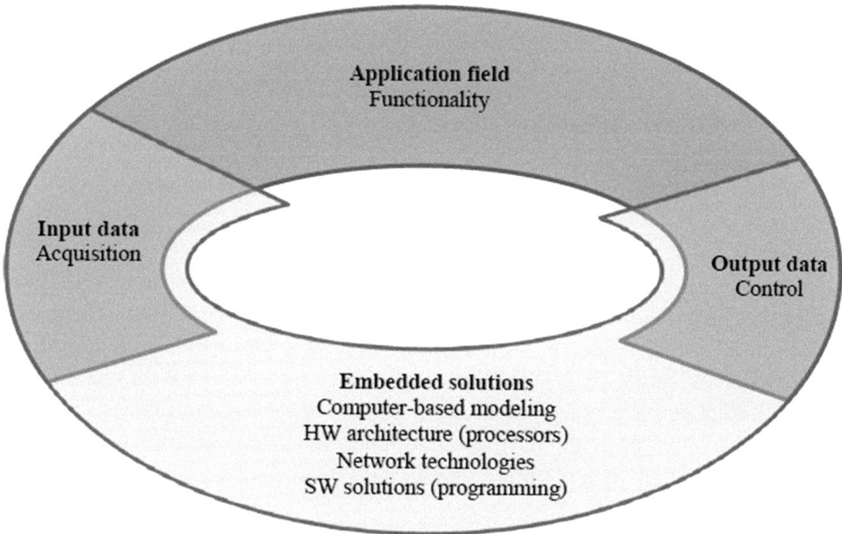

Fig. 1 Generic model of embedded solutions

depending on requirements and conditions, can vary from the simplest processor architecture (8-bit microcontroller for car window engine) up to complex multi-core solutions (system-on-chip for TV set-top boxes). In distributed applications (e.g. cloud computing) the corresponding networking concept has to be added. Finally, the required functionality should be implemented in SW programs. Variety of the used HW platforms covers programming from the lowest level (assembler and HW drivers) up to the highest level using complex SW tool environment (Windows, Linux, Android, iOS, etc.).

Finally, each embedded solution has to be tested and verified. The assessment of embedded solutions includes methodology and appropriate tools. Additional requirement is that the assessment of solutions provides some metrics which can be used in a feedback improvement loop.

The embedded engineering education should cover all these specific challenges of embedded solutions:

- Interfacing between application-based functionality and embedded-based solution. In the first line it is an interdisciplinary problem—how to formulate a problem from the application field using a formal (unified) embedded framework. Embedded system engineers need some basic understanding from targeted application fields (especially difficult is to formalize requirements from descriptive-oriented fields).
- Development-or-recombination approach. Often embedded solutions only require a recombination of already available units in a new context but in some other problems development of new units (knowledge or implementation) could be also required. Recognition of re-using opportunities requires a complex up-down learning process: how to abstract solution from a field and how to apply such abstracted approach in another field. Therefore, a usable formal abstraction of concrete problems and solutions should be covered by embedded engineering education.
- Balance between fundamental principles and practical usage of concrete platforms. Embedded engineering education faces a big dilemma—how to learn general principles which should be trained on variety of concrete platforms. Some solutions tend to establish a unified but extensible lab platform which can be used to train fundamental principles without introducing overload of new platforms in the labs.
- Assessment of embedded engineering education. The quality of achieved learning outcomes has to be measurable. The metrics should cover explicit knowledge units (e.g. processor architecture, operating systems, communication protocols) and skills (e.g. design, programming, application) achieved through lab trainings using concrete exemplar problems and available platforms. But it should also incorporate the development of soft skills (e.g. project management, teamwork, innovation capabilities) what could be difficult to cover by some formal metrics.

Covering these challenges we tried to give a formalized approach for embedded engineering education as a contribution to establish a structured embedded engineering study program.

4 Structured Approach for Embedded Engineering Education

In this paper we propose a structural approach for engineering curriculum using formal expressions of all identified items in the education process. Additionally, specifics of embedded engineering education will be considered.

The listed challenges in embedded engineering education (interdisciplinary problems, reusability of solutions, theory-practice dynamic balance and metrics for learning outcomes) specifically require a proactive role of industry in the education process. In distinction to the generally known education process, where an industry feedback comes after the finished education process, the embedded engineering education needs a continuous exchange during the education process. Therefore, the corresponding process flow should include two improvement loops as shown in Fig. 2.

The large loop, commonly for many education processes, includes the definition of needed engineering profiles from industry and provides educated engineers with required learning outputs from university. The needed profiles should consider actual market challenges expressed in skill requirements. Together with new

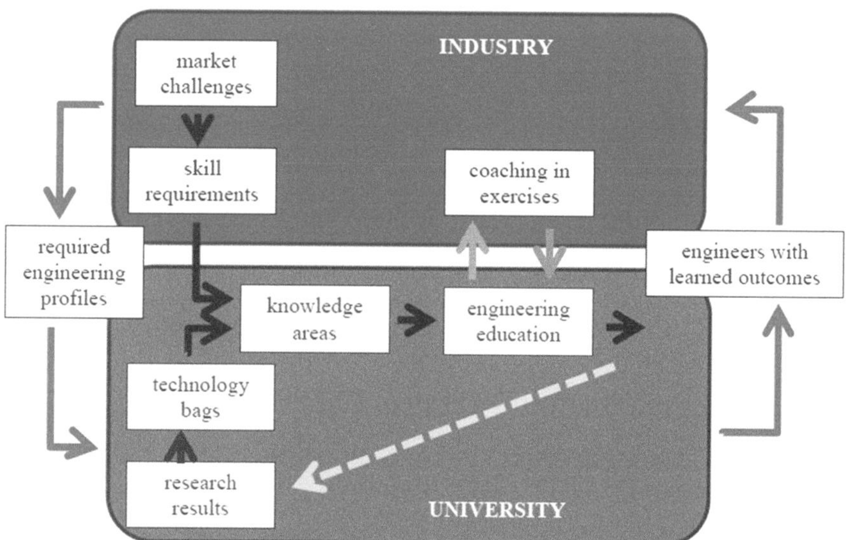

Fig. 2 Embedded engineering education process

developed technologies, coming from research, they define knowledge areas for the education process. Employment of the educated engineers in industry closes the education loop defining improvement needs and new requirements in a long term. Specifically for embedded engineering, a fast adaptation improving loop can be constructed involving experts from industry as the instructors in practical exercises during the education process. Additionally, part of educated engineers should be involved in further research bringing actual challenges and practical experience with modern technology.

The engineering education process should/can be structured using formal approaches. In this paper we propose some improvements in the formal approach of engineering education considering specifics of embedded engineering. The embedded engineering education process could be formally expressed using the known flow chart in Fig. 3.

Definition of an engineering study curriculum is based on the set of required skills and deviated knowledge areas. They are covered by study courses using lesson-exercise structure. The course structure has to fulfill two time integrity conditions: inter-dependence (sequential building up of knowledge) and limited student

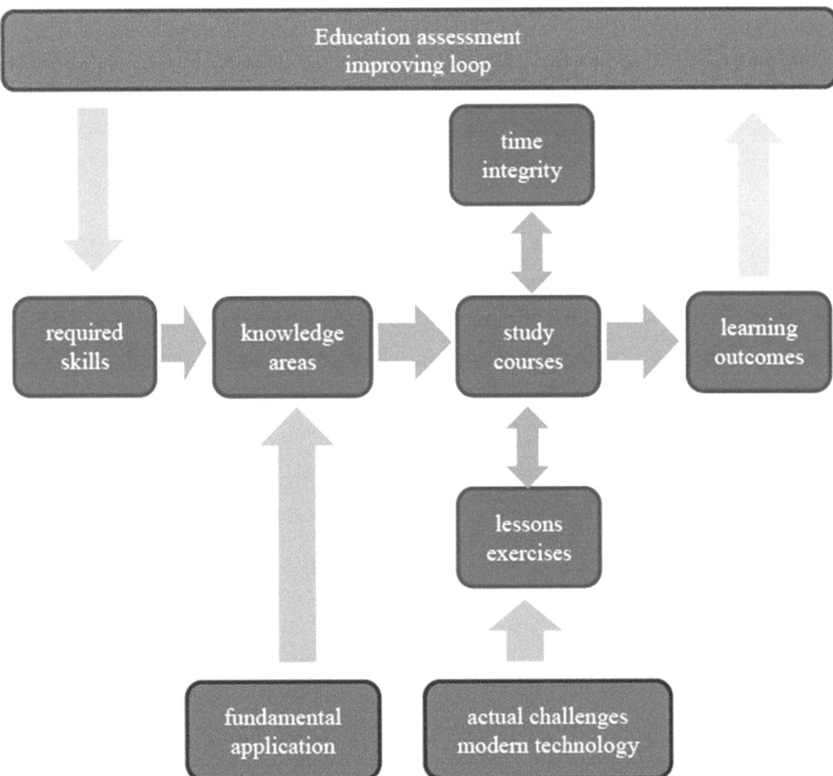

Fig. 3 Embedded engineering education process

load (study time limitations). The improving loop should be closed using some assessment metrics for continuous adaptations of the study curriculum. Specific requirements for embedded engineering education could be expressed in two additional steps: the knowledge areas should include basic understanding of application field (interdisciplinary approach and reusing approach) and lessons should be refreshed by actual challenges accompanied with exercises using modern technology. The set of measurable learning outcomes in education assessment should be extended with soft skills (project management, teamwork, innovation capacity).

Structural approach requires an appropriate numeric metrics. We are proposing a three-degree metrics:

F familiarity (understanding knowledge),
U usage (implementation capability), and
A acquired knowledge (deep understanding, capability of innovative reusing).

This metrics is used in definitions of all matching matrices, between skills and knowledge areas, between knowledge areas and study courses and between study courses and learning outcomes.

The required skills are coming from industry and often they are not well-structured. Identified in company operations (marketing, development, sales and consumer support) they could be generally sorted in three groups:

P predefined (established) engineering profiles like software engineering, computer engineering, mechanical engineering, chemistry etc.,
S specific knowledge like car engine, sensors, fluid theory etc., and
C concrete methods and tools like programming in Linux environment, CAN protocols, GNU editors, etc.

The list of knowledge areas are coming for university in a structured way. They are already identified and well-defined areas from research and education. Generally, they could be grouped in three main groups:

B basics covering high-abstract-level knowledge like mathematics, physics, etc.,
K fundamental knowledge like communications, computer architectures, programming, imaging, etc.,
E elective fields covering specific knowledge, methods and tools like network protocols, operating systems, compilers, etc.

The matching matrix between skills and knowledge areas contains dependencies expressed in the proposed metrics values. In practice, the skill list includes summarized needs coming from more companies.

A study curriculum is defined by study courses. They should be defined as learning-sustainable units respecting previous education experience. Courses are subdivided in lessons accompanied with appropriate practical exercises. The course construction is primary allocated at universities. But, regarding specifics of embedded engineering education, industry needs and experiences should be also incorporated in the definition of study courses. This influence of industrial experiences is especially important for incorporation of soft skills (teamwork, project management)

CDIO Syllabus v2.0	
Disciplinary knowledge • basic (math, science) • core engineering • advanced (methods and tools) • application context	**Interpersonal skills** • teamwork • communication • foreign languages
Personal and professional skills • analytic thinking • experimentation and discovery • system thinking • ethics	**Environmental skills** • social and environmental conditions • enterprise and business aspects • project management • implementation and operation aspects

Fig. 4 Framework for construction of study courses for embedded engineering

in engineering education. An appropriate way is the exercise construction based on problem-oriented and project-based approach [30]. The construction of study courses defines the whole engineering education framework. Therefore, the syllabus approach will be very helpful, as maintained in [2] and shown in Fig. 4.

The matching matrix between knowledge areas and study courses indicates coverage of a knowledge area in a study courses.

The achieved learning outcomes are indicated regarding lessons and exercises of all study courses. Standardly, each course yields more learning outcomes. This dependence is defined by a matching matrix between courses and learning outcomes.

Finally, the assessment matching matrix defines coverage of requested skills by achieved learning outcomes (0—not covered, 1—covered but not by requested degree and 2—covered by requested degree). This matrix could be used to check efficiency of the education process regarding the requested engineering profiles. The average of the assessment matching matrix values could be used as the global efficiency indicator of an education program for the given skill requirements.

Such structured approach in construction of engineering curricula additionally requires a sustainable timing order of defined courses, inter-dependences between the defined courses must be timely considered. Also, necessary efforts (hours per lesson, hours per exercise) must be aligned with a normal load of students (approximately 40 h per week). These frame conditions define the timing of a study program (semesters and courses).

The summarized visualization of the proposed approach is shown in Fig. 5.

Necessary interactions between university and industry in the embedded engineering education can be achieved using different models. We would propose one of them based on close collaborations between faculties and companies as shown in Fig. 6.

A company is interested for engineers with right profiles coming from faculty and sometimes for research results regarding open options on the market. The faculty is interested to focus real problems in their own education programs and to use modern technology in lab exercises. These interests on both sides could be the base for collaboration in engineering education processes. The most efficient method overcoming interfacing challenges between academy and industry is to put this interface conflict in one head—involving professors in company management

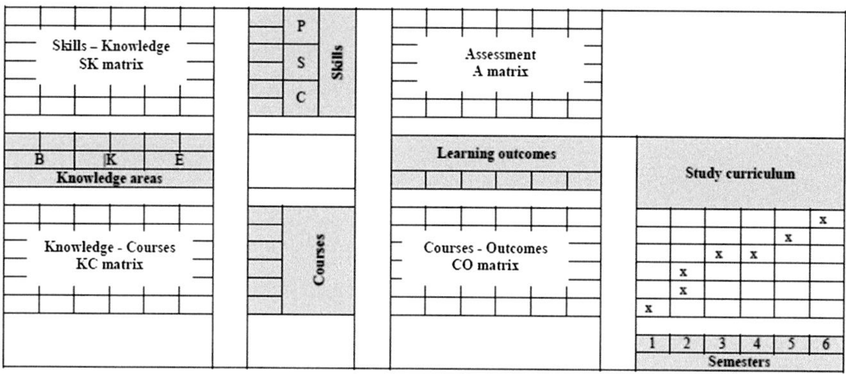

Fig. 5 Study curriculum scheme: matching matrices SK, KC and CO have F-U-A values (three-degree metrics), assessment matrix A has 0-1-2 values, study curriculum contains weekly hour-based lessons and exercises

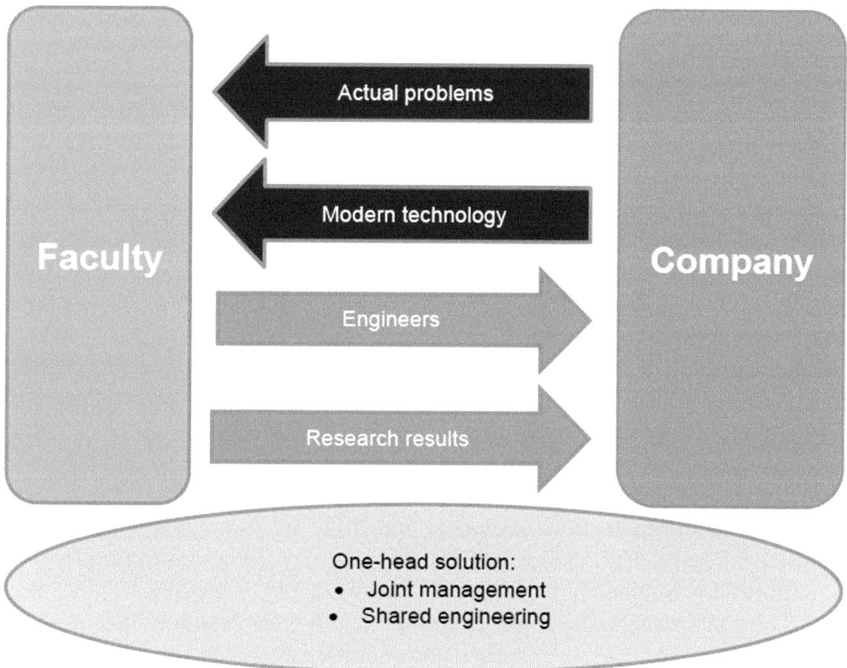

Fig. 6 Collaboration model between university and industry in engineering education

and including partially professional engineering team as instructors in the education process. P&L analysis shows in cases of dynamically developing fields that such approach reaches higher efficiency compared to standard engineering education and recruiting methods. The costs on the company side include provided lab equipment and partially allocated engineering resources. But, the company achieves a significant influence on the education process (shorter entrance trainings) and early access to further engineering potentials. In such kind of joint education the faculty has additionally to leverage between tight company interests and comprehensive engineering education. But, the actual challenges and soft skills (teamwork, project management) are directly included in the education process through shared engineering teams in labs.

5 Case Study—Unified Learning Platform

Faculty of Technical Sciences through its Department of Computing and Control Engineering established a strong link with industry with formation of the RT-RK company in 1991. After two decades, the company has grown to become a nationally recognized institute for computer based systems. Previously mentioned links between university and industry were established with professors participating in the management of the institute and employees of the institute with master education in engineering participating as laboratory assistants and demonstrators at the courses offered by the Department.

Study program for Computer engineering and communication, which is one of the study programs offered by the Department, has been developed over the years through its strong link with industry—students get involved with projects which are in line with the current industry demands as soon as their senior year of bachelor studies, helped in the lab by engineers from the institute. Master studies are organized in institute's laboratories where students get hands-on experience with working in team projects. After graduation, students are ready for job in any computer-engineering-related company or institute. This reduces the cost for preparation and education of the just-graduated engineer, which is a common requirement today. Selected few, who are interested in academic career, bring with them priceless experience from industry which is important for any educator in the computer engineering field because they are going to teach students, majority of which will pursue a career in industry.

In order to better organize education and make it more efficient, Faculty of Technical Sciences (FTN) joined RT-RK Institute and 7 other institutions to produce Embedded Engineering Learning Platform (E2LP) funded by EU FP7 program. The goal of the platform was to reduce the effort of preparing students at the start of each course which is common when multiple platforms and tools are used. Instead, students focus on concepts and content of the course, not losing time learning what to click and what to press in order to check their solution. In order to achieve this, platform was designed to support laboratory exercises in majority of the courses in the Computer engineering study program at FTN.

The following is a list of courses supported by the platform:

- Logic design of computer systems 1—digital system design
- Logic design of computer systems 2—computer system design
- Selected chapters of hardware design

- Real-time system software 1—parallel computing and compilers
- Real-time system software 2—operating systems
- Digital signal processing 1—basics of signals and systems
- Digital signal processing 2—architectures and algorithms for audio/image processing
- Computer networks 1—fundamentals of wired networks
- Computer networks 2—wireless networks
- Programming applications for mobile devices—Android

These courses span the majority of the curriculum in the second and third year of computer engineering. In the first year, students are introduced to the studies by learning basic disciplines—mathematics, physics, electrical engineering, computer architectures, programming and foreign language. In the second year they proceed with more math and programming courses, get some basics in control engineering, while being introduced to computer engineering in particular through digital system design and first part of the system software. The third year of this study program concentrates on computer engineering through all other courses listed above.

Benefit of E2LP platform is that in each course during the third year, students can start working on course-related material immediately in the theory and in the lab, since they already know the environment in which they will do the labs. The rest of this book will address this through the description of the platform and its extensions, as well as a detailed description of the methodology and evaluation of the pilot usage at FTN and other partner universities.

Senior year at the study program of computer engineering is organized as a project-based learning year in which students have less theoretical hours and more hours in the lab working on parts of the real projects from the institute. During their project work, while not formally integrated with the institute teams, students are working closely together with the engineers from the institute. Besides gaining theoretical and practical knowledge, they gain experience in job-related skills: teamwork, communication, research, deadline-driven work, etc. Fields that are studied during their fourth year in this manner are:

- Real-time software
- DSP architectures and algorithms—advanced
- Software for television and image processing
- Computer communications—advanced
- Optimization in DSP implementations
- Engineering of computer based systems
- Engineering of embedded systems

Tables 1, 2, 3, 4 and 5 give study curriculum scheme (from Fig. 5) for the computer engineering curriculum at FTN established in 2013.

After completing the final bachelor work, students graduate and should they continue with the master studies, they are completely integrated in the engineering teams in the institute, giving them access to the modern technologies and more industry experience.

Table 1 Skills/knowledge matrix in the case study

SK matrix		B			K			E		
		Math	Physics	Social	Computer systems	Software	System control	Networks	DSP	System SW
P	Computer engineer	F	F	F	A	U	F	A	A	A
	Computer science	U	F	F	U	A	F	U	U	U
S	Program	F	F	F	F	A	F	U	U	A
	Hardware design	F	F	F	A	U	F	U	U	U
	System design	F	F	F	A	U	A	U	U	U
C	C/C++	F	F	F	F	A	F	F	F	U
	Java	F	F	F	F	A	F	F	F	F
	VHDL	F	F	F	A	F	F	F	F	F
	Assembly	F	F	F	U	U	F	F	F	U
	Linux	F	F	F	F	U	F	F	F	A
	Android	F	F	F	F	U	F	F	F	A
	MATLAB	U	F	F	F	U	A	F	A	F

Table 2 Knowledge/courses matrix in the case study (transposed)

KC matrix	B			K			E		
	Math	Physics	Social	Computer systems	Software	System control	Networks	DSP	System SW
English language 1			F						
Mathematical analysis 1	F								
Discrete mathematics and linear algebra	F								
Programming languages and data structures					U				
English language 2			F						
Physics		F							
Fundamentals of electrical engineering		F							
Computer architecture				U					
Mathematical analysis 2	F								
Object-oriented programming					U				
Modeling and simulation						F			
Logic design of computer systems 1—digital system design				U					
Sociological aspects of technological development			F						
Probability and stochastic processes	F								
Control systems						F			
Operating systems									U
Real-time system software 1									U
Optimization methods						F			
Electronics		F							
Real-time system software 2									A
Digital signal processing 1								U	

(continued)

Table 2 (continued)

KC matrix	B			K			E		
	Math	Physics	Social	Computer systems	Software	System control	Networks	DSP	System SW
Computer networks 1—wired networks							U		
Logic design of computer systems 2—computer system design				U					
Selected chapters in hardware design				U					
Digital signal processing 2								U	
Computer networks 2—wireless networks							U		
Programming applications for mobile devices					U				
Real-time software 1					A				
Real-time software 2					A				
Architectures and algorithms of DSP 1								A	
Software in digital television and image processing								A	
Advanced computer networks							A		
Engineering of computer based systems				A					
Architectures and algorithms of DSP 2								A	
Embedded systems				A					
Practice				U	U		U	U	U
Thesis				A	A		A	A	A

Table 3 Courses/outcomes matrix in the case study

CO matrix	High level progr.	Low level progr.	Digital design	DSP progr.	Comm. SW design	App design	TV software	Broad edu
English language 1								F
Mathematical analysis 1								F
Discrete mathematics and linear algebra								F
Programming languages and data structures	U							
English language 2								F
Physics								F
Fundamentals of electrical engineering								F
Computer architecture		U						
Mathematical analysis 2								F
Object-oriented programming	U							
Modeling and simulation								F
Logic design of computer systems 1—digital system design			U					

(continued)

Table 3 (continued)

CO matrix	High level progr.	Low level progr.	Digital design	DSP progr.	Comm. SW design	App design	TV software	Broad edu
Sociological aspects of technological development								F
Probability and stochastic processes								F
Control systems								F
Operating systems		U						
Real-time system software 1		U						
Optimization methods								F
Electronics								F
Real-time system software 2		A						
Digital signal processing 1				U				
Computer networks 1—wired networks					U			
Logic design of computer sys-tems 2—computer system design		U	U					
Selected chapters in hardware design			U					
Digital signal processing 2				U			U	

(continued)

Table 3 (continued)

CO matrix	High level progr.	Low level progr.	Digital design	DSP progr.	Comm. SW design	App design	TV software	Broad edu
Computer networks 2—wireless networks					U			
Programming applications for mobile devices						U	U	
Real-time software 1	A							
Real-time software 2	A							
Architectures and algorithms of DSP 1				A				
Software in digital television and image processing							A	
Advanced computer networks					A			
Engineering of computer based systems	A	A						
Architectures and algorithms of DSP 2				A			A	
Embedded systems	A	A						
Practice	U	U	U	U	U	U	U	
Thesis	A	A	A	A	A	A	A	

Table 4 Assessment matrix in the case study

A matrix		High level program.	Low level program.	Digital design	DSP program.	Communication software design	Application design	TV software
P	Computer engineer	1	2	2	2	2	1	2
	Computer science	2	1	1	1	1	1	1
S	Program	2	2	1	2	2	2	2
	Hardware design	0	1	2	0	0	0	0
	System design	1	1	1	1	1	0	0
C	C/C++	2	2	1	2	2	2	2
	Java	2	0	0	1	1	2	2
	VHDL	0	0	2	0	0	0	0
	Assembly	1	2	1	2	1	0	1
	Linux	1	2	0	1	1	0	1
	Android	1	2	0	1	1	1	1
	MATLAB	0	0	0	2	0	0	0

Table 5 Study curriculum of computer engineering at FTN

Semester	1	2	3	4	5	6	7	8
English language 1	3 + 0							
Mathematical analysis 1	4 + 4							
Discrete mathematics and linear algebra	4 + 4							
Programming languages and data structures	4 + 4							
English language 2		3 + 0						
Physics		4 + 4						
Fundamentals of electrical engineering		4 + 4						
Computer architecture		4 + 4						
Mathematical analysis 2			4 + 4					
Object-oriented programming			4 + 4					
Modeling and simulation			4 + 4					
Logic design of computer systems 1—digital system design			3 + 3					
Sociological aspects of technological development				2 + 0				
Probability and stochastic processes				2 + 2				
Control systems				4 + 4				
Operating systems				4 + 4				
Real-time system software 1				3 + 3				
Optimization methods					4 + 4			
Electronics					4 + 4			
Real-time system software 2					2 + 4			
Digital signal processing 1					2 + 2			
Computer networks 1—wired networks					2 + 2			

(continued)

Table 5 (continued)

Semester	1	2	3	4	5	6	7	8
Logic design of computer systems 2—computer system design						4 + 4		
Selected chapters in hardware design						3 + 3		
Digital signal processing 2						2 + 2		
Computer networks 2—wireless networks						2 + 2		
Programming applications for mobile devices						4 + 4		
Real-time software 1							2 + 2	
Real-time software 2							2 + 2	
Architectures and algorithms of DSP 1							4 + 3	
Software in digital television and image processing							4 + 3	
Advanced computer networks							3 + 2	
Practice							0 + 3	
Engineering of computer based systems								3 + 3
Architectures and algorithms of DSP 2								3 + 3
Embedded systems								4 + 4
Thesis								0 + 9

6 Conclusions

Evident importance of the embedded engineering education requires new innovative concepts to create and to implement the corresponding study programs. The main challenges in these considerations include:

- Interdisciplinary approach having in mind set of broad relevant applications which require knowledge from different fields,
- Deep familiarity with available methodology and tools in computer engineering, and
- High development dynamics regarding new technology and methods.

In this paper a contribution to the structural approach in engineering education is proposed based on matching matrices between:

- Required skills and knowledge areas,
- Knowledge areas and learning courses,
- Learning courses and achieved outcomes.

This structure is accompanied with an assessment matrix between the achieved outcomes and the required skills providing the improvement loop of the education process. Finally, the set of learning courses should define a study program considering education limitations (time and space). Additionally, a model how to involve industry in the education process is proposed considering continuous technology development.

The proposed concept is illustrated in a case study—the study program of computer engineering at the University of Novi Sad.

References

1. Koopman, P.: Challenges in embedded systems research and education. http://users.ece.cmu.edu/~koopman/embedded/challenges.pdf (1995). Accessed 20 July 2015
2. Improving Engineering Education, CDIO common program from MIT, KTH, LiU, Chalmers. MIT (2000)
3. CDIO—annual report 2003, CDIO program master roadmap (2003)
4. Sztipanovits, J., Biswas, G., Frampton, K., Gokhale, A., Howard, L., Karsai, G., Koo, T.J., Koutsoukos, X., Schmidt, D.C.: Introducing Embedded Software and System Education and Advanced Learning Technology in an Engineering Curriculum. Vanderbilt University Initiative
5. ACM/IEEE-CS Joint Task Force on Computing Curricula, Computer Science Curricula 2013, ACM/IEEE Press, Technical Report, Dec 2013, http://dx.doi.org/10.1145/2534860. Accessed 20 July 2015
6. ACM/IEEE-CS joint task force on computing curricula. In: Computer Engineering Curricula 2004 (2004)
7. Winzker, M., Schwandt, A.: Teaching embedded system concept for technological literacy. In: IEEE International Conference on Microelectronic System Education 2009, pp. 89–92
8. Marijan, D., Zlokolica, V., Teslic, N., Pekovic, V., Tekcan, T.: Automatic functional TV set failure detection system. IEEE Trans. Consum. Electron. **56**, 125–133 (2010)

9. Cunha, J.C., Amaro, J.P., Marques, L.: A joint academy-industry initiative for the development of an engineering program. In: 1st International Conference of the Portuguese Society for Engineering Education (2013)
10. Pokkiyarath, M., Raman, R., Achuthan, K., Jayaraman, B.: Preparing global engineers: USA-India academia & industry led approach. In: IEEE (2014)
11. Ktoridou, D., Eteoleous, N.: Engineering education: time to reform the fragmented, content-overloaded curricula context? In: IEEE Global Engineering Education Conference EDUCON, 2014, pp. 377–380
12. Karsai, G., Massacci, F., Osterweil, L.J., Schieferdecker, I.: Evolving embedded systems. In: IEEE Computer, pp. 34–40 (2010)
13. MacBride, G., Hayward, E.L., Hayward, G., Spencer, E., Ekevall, E., Magill, J., Bryce, A.C., Stimpson, B.: Engineering the future: embedded engineering permanently across the school-university interface. IEEE Trans. Educ. **53**(1), 120–127 (2010)
14. Bruce, R., Brock, D., Reiser, S.: Teaching programming using embedded systems. In: IEEE (2013)
15. Banavar, M.K., Rajan, D., Strom, A., Spanias, P., Zhang, X., Braun, H., Spanias, A.: Embedding Android signal processing apps in high school math class—an RET project. In: IEEE (2014)
16. Sangiovanni-Vincentelli, A., Martin, G.: Platform-based design and software design for embedded systems. IEEE Des Test Comput **18**(6), 23–33
17. Kastelan, I., Lopez Benito, J.R., Artetxe Gonzalez, E., Piwinski, J., Barak, M., Temerinac, M.: E2LP: a unified embedded engineering learning platform. Elsevier Microprocess Microsyst. **38**(8 Part B), 933–946
18. Koopman, P.: Better Embedded System Software, Drumnadrochit Education (2010)
19. Nakutis, Z., Saunoris, M.: Challenges of embedded systems teaching in electronic engineering studies. J. Electron. Electr. Eng. **6**(102), 83–86 (2010)
20. Reichenbach, M., Pfundt, B., Fay, D.: Designing and manufacturing of real embedded multi-core CPUs: a holistic teaching approach in computer architecture. In: IEEE (2014)
21. Stettina, C.J., Zhou, Z., Baeck, T., Katzy, B.: Academic education of software engineering practices: towards planning and improving capstone courses based upon intensive coaching and team routines. In: CSEE&T 2013, pp. 169–178 (2013)
22. Zhan, J., Stoimenov, N., Oyang, J., Thiele, L., Narayanan, V., Xie, Y.: Designing energy-efficient NoC for real-time embedded systems through slack optimization. In: 50th Annual Design Automation Conference (2013)
23. Jayasinghe, U., Dharmaratne, A., Atukorale, A.: Education system versus traditional education system. In: 12th International Conference on Remote Engineering and Virtual Instrumentation 2015, pp. 131–135 (2015)
24. Blank, M., Brunner, S., Fuhrmann, T., Meier, H., Niemetz, M.: Embedded Linux in engineering education. In: IEEE EDUCON 2015, pp. 145–150 (2015)
25. Sandanayake, T.C., Madurapperuma, A.P., Davis, D.: Affective e-learning model for recognizing learner emotions. Int. J. Inf. Educ. Technol. **1**, 315–320 (2011)
26. Luthon, F., Larroque, B.: LaboREM—a remote laboratory for game-like training in electronics. IEEE Trans. Learn. Technol., 1–12 (2014)
27. Trenas, M.A., Ramos, J., Gutierrez, E.D., Romero, S., Corbera, F.: Use of new Moodle module for improving the teaching of a basic course on computer architecture. IEEE Trans. Educ. **54**(2), 222–228 (2011)
28. Guimares, E.G., Cardozo, E., Moraes, H., Coelho, P.R.: Design and implementation issues for modern remote laboratories. IEEE Trans. Learn. Technol. **4**(2), 149–161
29. Lopez Benito, J.R., Artetxe Gonzalez, E., Anastassova, M., Souvestre, F.: Engaging computer engineering students with an augmented reality software for laboratory exercises. In: IEEE Frontiers in Education Conference FIE, 2014, pp. 1–4 (2014)

30. Lehman, M., Christensen, P., Du, X., Thrane, M.: Problem-oriented and project-based learning (POPBL) as an innovative strategy for sustainable development in engineering education. Eur. J. Eng. Educ. **33**(3), 283–295 (2008)
31. Dogmus, Z., Erdem, E., Patoglu, V.: REACT: an interactive educational tool for AI planning for robotics. IEEE Trans. Educ. **58**(1), 15–24 (2015)
32. Borri, C., Guberti, E., Quadrado, J.C.: Quality assurance of engineering education worldwide. In: IEEE International Conference on Interactive Collaborative Learning, 2014, pp. 346–351 (2014)
33. Raju, P.K., Sankar, C.S., Halpin, G., Halpin, G., Good, J.: Evaluation of an engineering education coursware across different campuses. In: 30th ASEE/IEEE Frontiers in Education Conference, 2000, T4B-11 (2000)
34. Ristov, S., Ackovska, N., Kirandzirska, V.: Positive experience of the project gamification in the microprocessors and microcontrollers course. In: IEEE Global Engineering Education Conference EDUCON, 2015, pp. 511–517 (2015)
35. Persad, K., Ringis, D., Radix, C.: Leveraging student motivation in engineering skills acquisition. In: IEEE (2014)
36. CDIO initiative. http://www.cdio.org/. Accessed 20 July 2015
37. Google Teacher Academy. https://www.google.com/edu/resources/programs/google-teacher-academy/. Accessed 20 July 2015
38. Intel Academy. https://engage.intel.com/community/edacademy. Accessed 20 July 2015
39. Microsoft IT Academy. https://www.microsoft.com/en-us/education/it-academy/default.aspx#fbid=ry-elMZh5_s. Accessed 20 July 2015

Unified Learning Platform for Embedded Engineering

Ivan Kastelan, Nikola Teslic and Miodrag Temerinac

Abstract This paper presents the hardware and software architecture of a unified embedded engineering learning platform. The platform consists of a base board with FPGA and extension boards with microprocessors. The goal of the platform is to support laboratory exercises in the entire embedded engineering curriculum, reducing the overhead in giving tutorials to students and concentrating on the core material from each course. The platform has been evaluated with students and teachers in courses of the computer engineering curriculum at Faculty of Technical Sciences. The paper gives impressions from the pilot generation of students using the platform.

Keywords Embedded engineering · Learning platform · Education

1 Introduction

The success of the education of future engineers of electrical engineering and computer science largely depends on the quality of laboratory work during their studies. Through laboratory work it is possible to enhance the role of students from passive listeners to active participants, which further encourages them to participate in the process of learning [1]. Knowledge gained through practical experience in the laboratory is shown as deeper and more permanent.

I. Kastelan (✉) · M. Temerinac
Faculty of Technical Sciences, University of Novi Sad, Trg Dositeja Obradovica 6, 21000 Novi Sad, Serbia
e-mail: ivan.kastelan@rt-rk.uns.ac.rs

N. Teslic
RT-RK Institute for Computer Based Systems, Narodnog Fronta 23a, 21000 Novi Sad, Serbia

R. Szewczyk et al. (eds.), *Embedded Engineering Education*, Advances in Intelligent Systems and Computing 421, DOI 10.1007/978-3-319-27540-6_2

Industry recognizes the growing need for highly educated personnel in the field of electrical engineering, computer engineering technology and embedded systems [2–5]. In response, many electrical engineering programs increasingly put an emphasis on learning in the field of embedded systems, mainly relying on the increasing number of laboratory-based courses that largely follow the principle of so-called active learning. The dynamics needed to satisfy the needs of the industry places teachers responsible for the design of laboratory environments required to conduct courses under considerable pressure. The situation is further complicated by the fact that, due to the dynamic development of the field, laboratory environment needs to be regularly renewed. Finally, it is desirable that the program of instruction uses different and often inconsistent laboratory environments and learning platforms.

Quality and efficiency of laboratory work in the field suffers due to the additional time and effort required to acquaint students with the hardware platforms and software tools for each course. Because of this, especially at the beginning of the course, it is difficult for students to divert attention from the used tools to the fundamental principles that constitute the main contribution of the course.

Engineering education is one of the most dynamic types of education because of new technologies in the field which occur rapidly. Therefore, it is necessary to constantly be up to date with the development of new technologies and students need to learn using the latest teaching and laboratory resources.

To overcome the above problems, the Faculty of Technical Sciences in cooperation with another 8 institutions from 7 European countries developed a universal platform for laboratory exercises in the field of embedded computer systems—E2LP platform [6]. It is a platform that supports laboratory work in the majority of cases required for the training of engineers in this field and thereby accelerates their knowledge by reducing unnecessary consumption of time to become familiar with a variety of platforms. Wide range of courses that this platform supports is enabled by the development of extension boards that connect to the base board.

Based on previous observations and industrial experience in the field of embedded systems, a unified modular platform for learning was developed. The platform aimed to increase the efficiency of lab-based courses. It is based on the generation of field programmable gate array (FPGA) manufactured by Xilinx—Spartan-6. The platform is designed to cover all aspects of embedded engineering including: (1) the design of digital systems, (2) the design of computer systems, (3) signal processing, (4) computer networks and interfaces, (5) system integration [7]. Maximizing the efficiency of lab-based courses is based on the idea of using the implemented unified platform throughout the entire curriculum. The main contribution is the effective education of future engineers capable to face the current challenges in the field of embedded engineering and their applications in real time. Software for the platform was developed with aims to help students use the platform more easily and inform the students about the platform operation at a given moment (the value of the signals, clock frequency, etc.).

This paper describes the basic elements of the E2LP platform:

- Base board with FPGA,
- Extension boards based on digital signal processor (DSP) and ARM processors and
- Software for the platform.

This paper is organized as follows: Sect. 2 presents the architecture of E2LP base board. Section 3 describes extension boards. Software is presented in Sect. 4, while Sect. 5 gives first impressions from the pilot generation using the platform. Section 6 gives concluding remarks and plans for future improvements.

2 E2LP Base Board

The E2LP Base Board is based on an FPGA in the Xilinx Spartan-6 family. In conjunction with the Xilinx ISE and XPS programming environments, it is a complete solution for the design of digital and computer systems with FPGA components (Fig. 1).

The aforementioned Spartan-6 FPGA integrated circuit is surrounded by a set of peripherals that can be used to generate complex systems. Figure 2 shows a block diagram of the development board with all available peripherals and connections between individual components marked.

The E2LP Base Board performs the following functions:

- based on FPGA, provides the central point of the E2LP platform on which all other parts are connected;
- supplies power for the whole E2LP platform;
- controls programming the FPGA and central processing units (CPUs) on extension boards;
- provides a basic user interface;
- provides storage, multimedia and communication interfaces for the platform;
- provides the platform for digital system design;
- provides test points for debugging.

The key building modules of the E2LP Base Board (Fig. 2) are:

- Xilinx Spartan-6 FPGA,
- ARM-based control processor,
- Mezzanine connector to extension board (Xilinx FMC LPC standard),
- DDR2, flash and multimedia card memory,
- user interface (8 switches, 6 buttons, 8 LEDs, alphanumeric LCD screen),
- snapwire connector,
- CVBS video encoder and decoder,
- video output (VGA, HDMI),
- audio sub-system,
- communication interfaces (USB, Ethernet, RS232 and Infra-red).

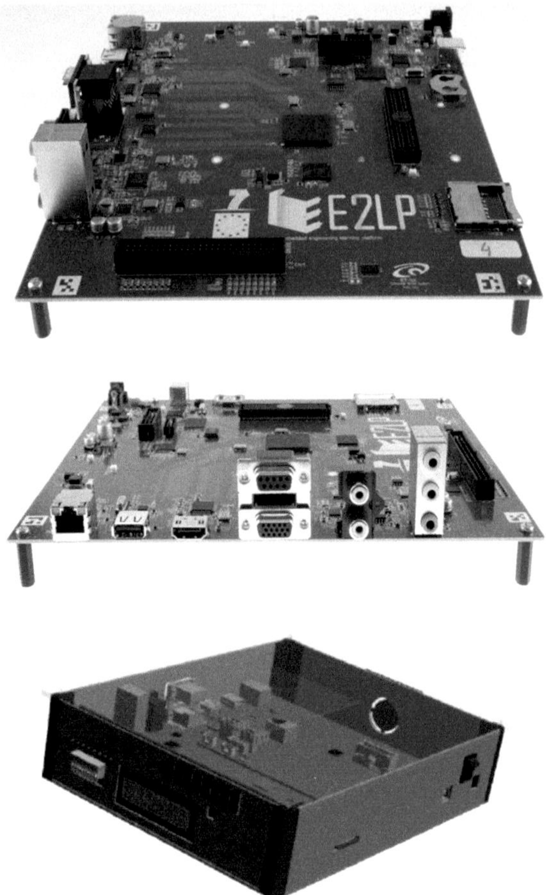

Fig. 1 E2LP Base Board top view (*up*), side view with some of the multimedia and communication interfaces connected to FPGA (*middle*) and 3D view (*bottom*)

The E2LP Base Board, as presented in this paper, together with its extension boards, is working in fully satisfying the main requirement of the E2LP platform—to be used in the complete embedded engineering curriculum and significantly reduce the overhead in engineering education. Implementation of the extension boards whose mechanical requirements are dependent on the Base Board will be explained in the next section.

FPGA component can be configured directly using the Joint Test Action Group (JTAG) coupling system, or indirectly using a dedicated flash memory on board. Flash component that is on the board gives the possibility to use up to four different revisions of the code for FPGA configuration. Personal computer (PC) connects via the appropriate cable (e.g. Xilinx Parallel Cable IV or Xilinx Platform Cable USB) for programming via JTAG coupling system in both cases: for direct FPGA configuration and for programming flash memory.

Due to the high cost of Xilinx's equipment for FPGA configuration via JTAG interface, a custom configuration via USB interface was developed for the E2LP

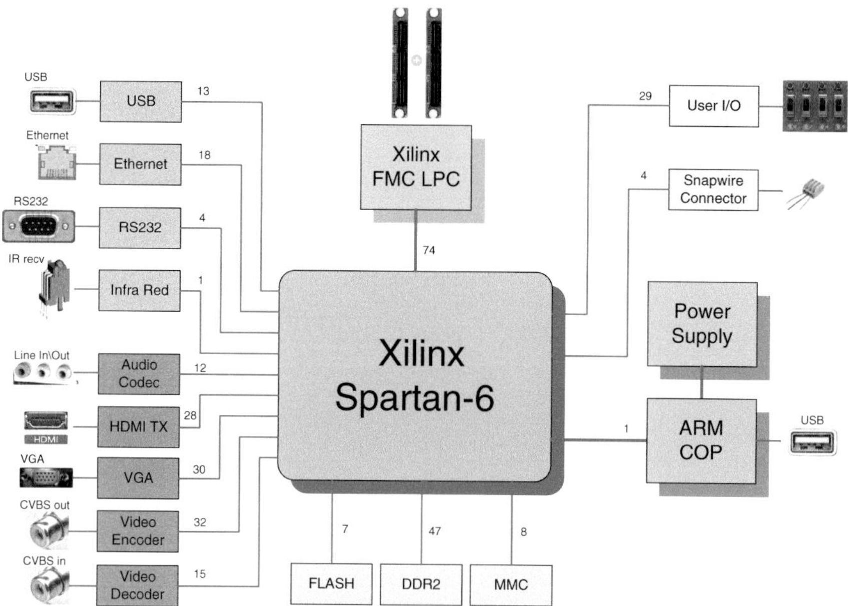

Fig. 2 Architecture of E2LP Base Board

platform. This innovative way of configuration is using a standard USB cable and is supported by specially developed software for E2LP platform. The custom method of FPGA configuration can bring significant savings to potential users, keeping in mind the price of standard solutions for programming (aforementioned configuration cables).

Complex programmable logic device (CPLD) is used for the selection of one of the three options of FPGA configuration. The presence of extension boards, as well as the desire of the ARM on Base Board to program the FPGA, is managed by the CPLD integrated circuit which can be configured in one of three configurations shown in Fig. 3:

1. Chain of FPGA, CPLD and Flash—configuration is performed via the JTAG Platform Cable.
2. Chain of FPGA, CPLD, Flash and ARM—ARM CPU is used to configure FPGA.
3. Chain of FPGA, CPLD, Flash, ARM and extension board—extension board is attached and JTAG-configurable parts can be configured on it.

Bearing in mind the need for constant improvement of the laboratory environment, providing an elegant way to extend the functionality of the platform was taken into consideration. In order to achieve that, connection with extension boards is implemented using the connector with small capacitance and high speed to ensure signal integrity even at high speeds of data transfer.

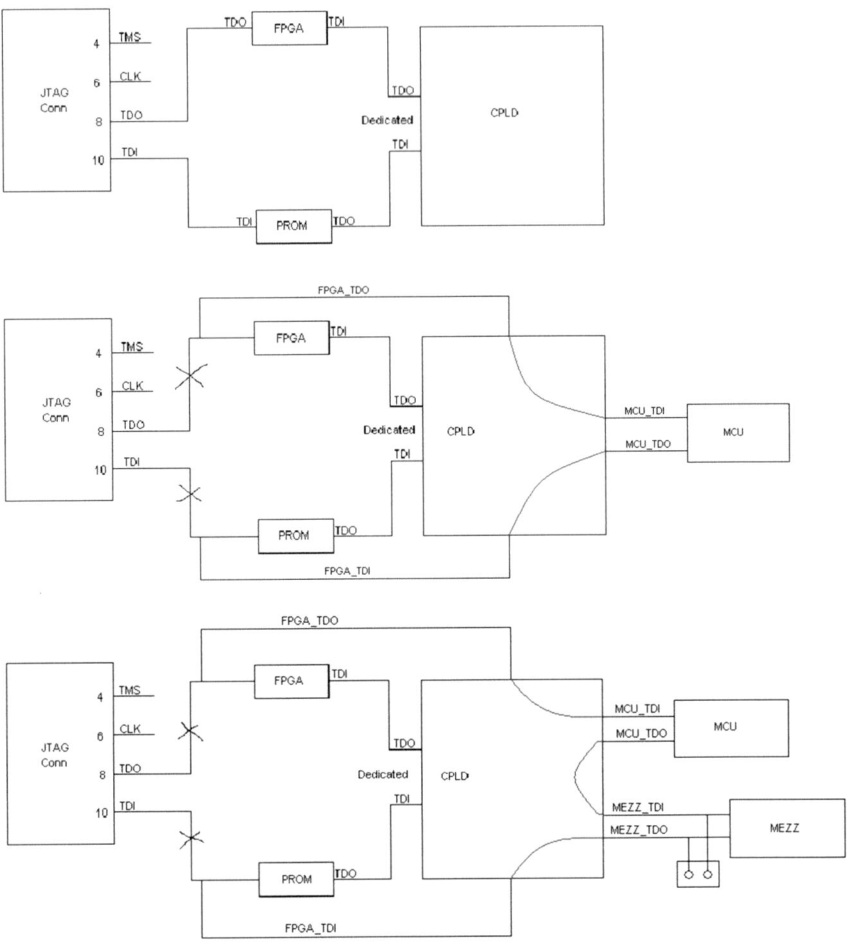

Fig. 3 Three versions of JTAG chain for FPGA configuration: using the Platform Cable (*upper*), using the USB connection to PC application (*middle*) and using USB with extension board support (*lower*)

In order to make the solution as general as possible, Xilinx FMC LPC connector standard was used. This connector provides 80 lines connected to the FPGA pins, of which 68 are general purpose input/outputs through which the indirect access to all peripherals on the board is enabled. In addition, the connector has access to the JTAG chain, I2C bus and user-programmable generators that allow the synthesis of clock signals in a wide frequency range.

The next section describes the implemented extension boards which connect to the E2LP Base Board via the FMC LPC interface.

3 Extension Boards

Three extension boards were implemented for the E2LP Base Board:

- Extension board based on ARMADA processor
- Extension board based on ARM LPC processor
- Proto-board extension

3.1 ARMADA Extension Board

One of the extension boards of the E2LP platform was designed to support laboratory exercises in the field of embedded computer systems based on the Marvell processor 88DE3108, known as ARMADA 1500. This processor is connected to the RAM memory in the form of two DDR3 modules with a capacity of 4 Gb (256 M $\times$ 16), i.e. 8 Gb in total. The modules are connected to the so-called "fly by" topology.

The extension board based on Marvell ARMADA 1500 has the following functions:

- based on ARM processor, provides the extension to the E2LP platform suitable for highly sophisticated signal processing and execution of real-time software;
- connects to the E2LP base board via Mezzanine connection;
- connects to the exterior with USB, LAN and HDMI interfaces;
- provides the extension to the E2LP platform suitable for implementing laboratory exercises in the field of digital signal processing, real-time system software, computer networks and system integration.
- provides test points for debugging.

Two HDMI interfaces allow connecting devices and displaying images on monitors and/or TV. To achieve an elegant way of extending the functionality of the platform, expansion ports are implemented using Xilinx FMC connector with small capacitance and large speed, in order to ensure signal integrity even at high speeds of data transfer.

The block diagram (Fig. 4) gives a high level overview of the E2LP extension board based on Marvell ARMADA 1500 processor.

ARMADA extension board is managed by Android operating system. It is chosen due to its popularity as it is recording a growing presence in the market of mobile devices. It is estimated that Android OS is currently used by hundreds of millions of mobile devices around the world. This board uses version 4.2.2 (Jelly Bean) with built-in support for Google TV (ver. 82210). Source code for Android is available under the "free and open" license for use. Android SDK allows students to make application through its applicative interface (API) as well as the necessary tools for developing Android applications in the Java programming language.

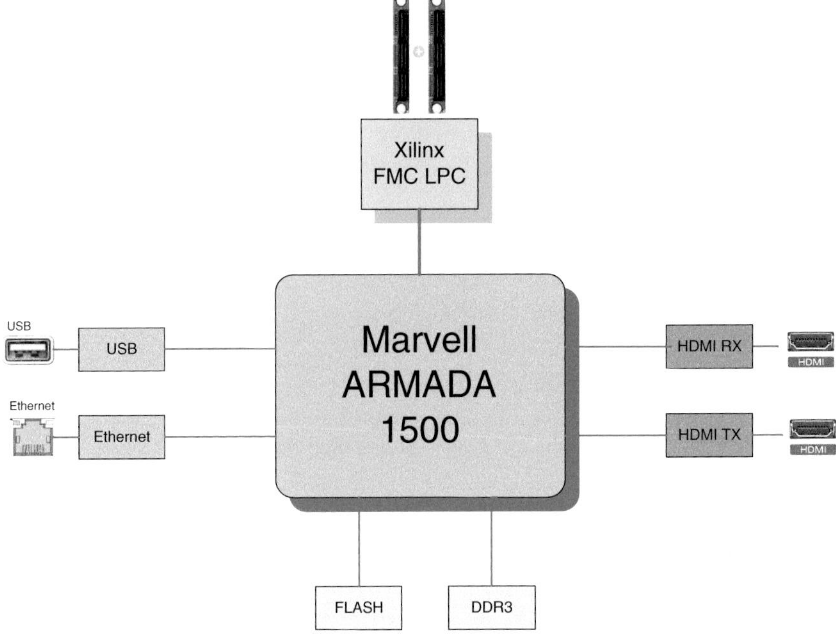

Fig. 4 E2LP extension board with ARMADA 1500 processor

This extension board allows implementation of the laboratory exercises in the following subject categories:

- 1-D digital signal processing
- 2-D digital signal processing
- computer networks and communications
- system integration
- system software
- Android development

The extension board can use all the resources from the E2LP Base Board and in addition to that, it can use its own USB, HDMI, LAN, Flash and DDR3. ARMADA can be programmed via JTAG interface through the Base Board and connected to PC for debugging and execution.

Also, E2LP extension board based on Marvell ARMADA 1500 can be used independent. It has possibility to be powered externally through JST's header 1×4 2 mm vertical connector (12 and 3.3 V voltage level is needed).

3.2 NXP LPC ARM Extension Board

In order to support laboratory exercises in courses for which a complex processor and operating system are not required, and to support low-level programming, the second extension board was developed with the basic ARM microcontroller. It contains:

- LPC2364 microcontroller
- DS18S20 High-Precision 1-Wire Digital Thermometer
- LM386 Low Voltage Audio Power Amplifier
- BMA250 digital accelerometer and I2C
- Snapwire connector with 8-pins
- Push-button switches, rotary encoder and LEDs
- TJA1040 High speed CAN transceiver.

The block diagram (Fig. 5) gives a high level overview of this extension board. The following are specifications of this extension board:

- 10-bit ADC with input multiplexing among 6 pins.
- 10-bit DAC.
- Four general purpose timers/counters with a total of 8 capture inputs and 10 compare outputs. Each timer block has an external count input.
- One PWM/timer block with support for three-phase motor control. The PWM has two external count inputs.
- Real-Time Clock (RTC) with separate power pin, clock source can be the RTC oscillator or the APB clock.

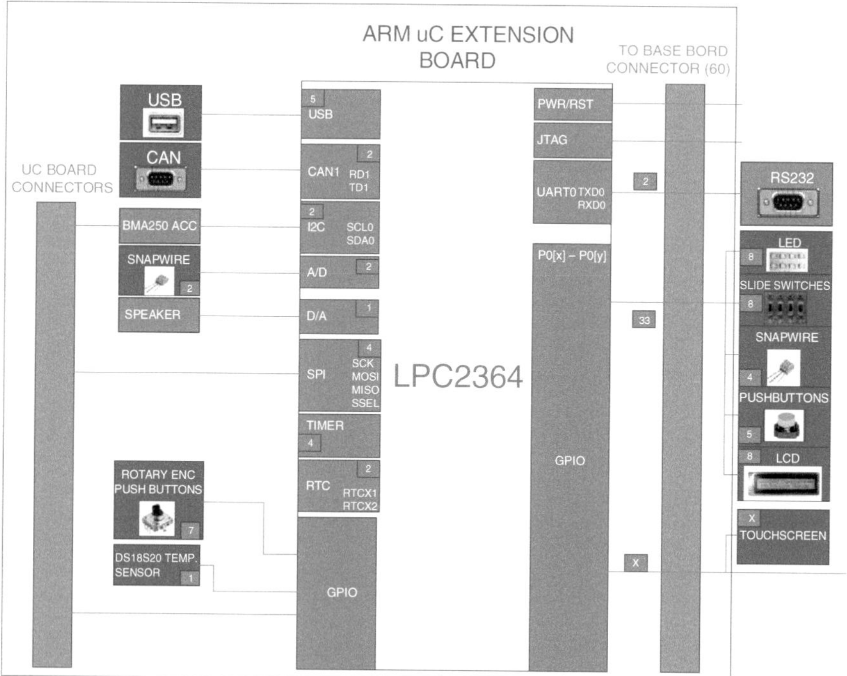

Fig. 5 E2LP extension board with NXP LPC ARM processor

- 2 kB SRAM powered from the RTC power pin, allowing data to be stored when the rest of the chip is powered off.
- WatchDog Timer (WDT). The WDT can be clocked from the internal RC oscillator, the RTC oscillator, or the APB clock.
- Standard ARM test/debug interface for compatibility with existing tools.
- Emulation trace module supports real-time trace.
- Single 3.3 V power supply (3.0–3.6 V).
- Four reduced power modes: idle, sleep, power-down, and deep power-down.
- On-chip crystal oscillator with an operating range of 1–24 MHz.
- 4 MHz internal RC oscillator trimmed to 1 % accuracy that can optionally be used as the system clock. When used as the CPU clock, does not allow CAN and USB to run.

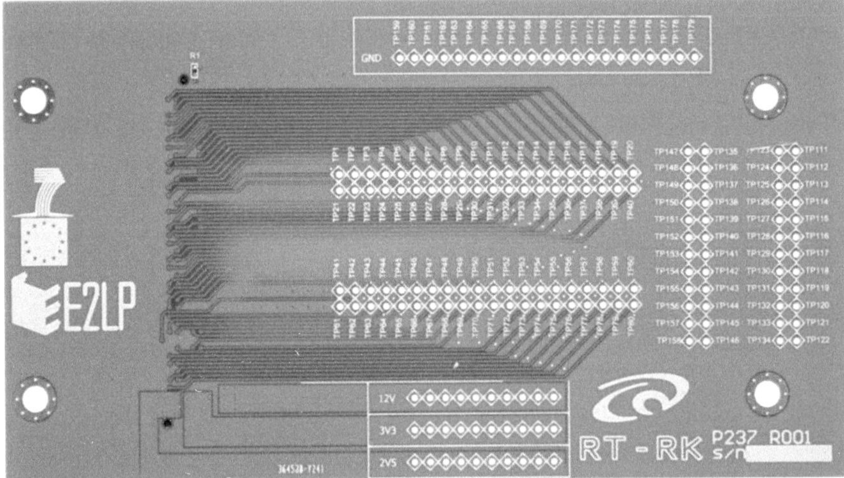

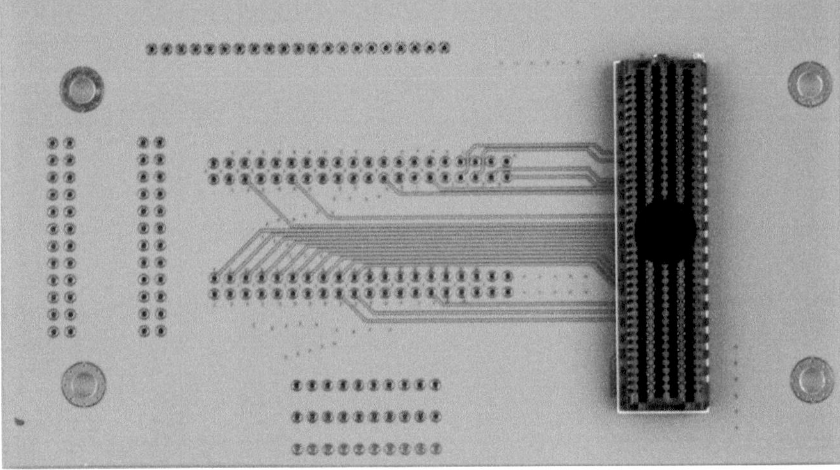

Fig. 6 E2LP proto-board extension—*top* and *bottom* view

This extension board allows implementation of the laboratory exercises in the following subject categories:

- Basic low-level programming
- Computer architecture
- Simple high-level programming (C language)
- Control engineering
- Industrial software

3.3 Proto-Board Extension

The third extension board is a simple proto-board (Fig. 6) which provides open access to every general-purpose pin on the FMC mezzanine connection to FPGA on Base Board. This board is useful for practicing basic electronics and building devices which connect directly to FPGA. Since the pins connect directly to FPGA, it is user's responsibility to adhere to the FPGA's electrical standards.

4 E2LP Software

Software for E2LP base board system is divided in two parts. The first part is software developed for housekeeping microcontroller (PHILLIPS LPC214x). The second part is software developed for PC working station. Communication between those two parts is established via USB connection.

The software for housekeeping microcontroller (Fig. 7) is in charge of the following tasks:

- Monitoring of board's power supply levels via AD converters which are part of microcontroller unit (MCU)
- Checking if FPGA is configured
- USB HID interface bring-up
- Sending monitored data to working station side through USB connection.

Program is stored in the flash ROM of the LPC214x and is started upon board's power-up or after resetting MCU.

The core of the MCU's application is USB HID kernel. All monitoring data are read periodically form the USB interrupt service routine which is invoked every millisecond.

Application for the PC working station is designed to enable users to monitor base board parameters:

- power supply levels
- configuration state of FPGA on Base Board.

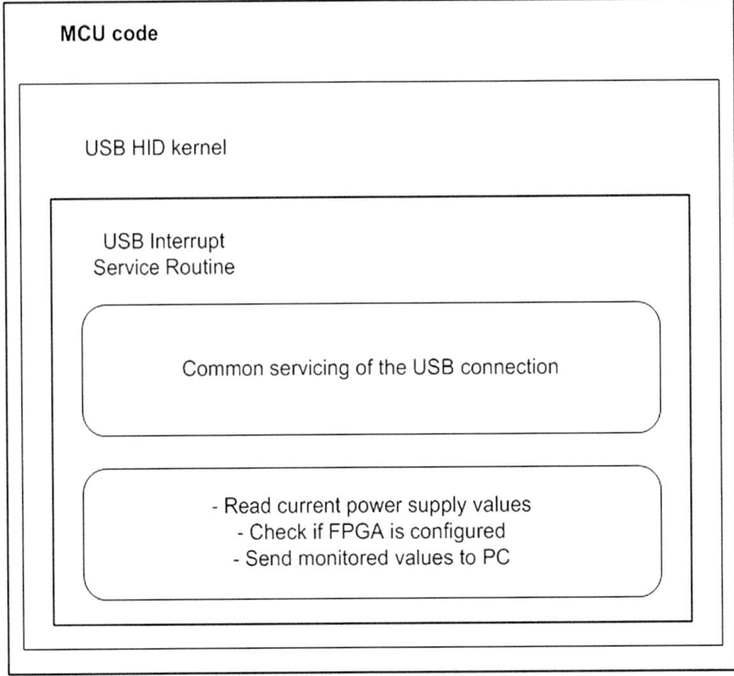

Fig. 7 Architecture of MCU software

It collects information from the board's housekeeping MCU via USB connection and displays them to the user.

Using the developed E2LP configuration utility, the user can:

- connect its PC via USB to the Base Board and control it,
- check the state of voltages on Base Board,
- control whether the application is connected to the JTAG chain,
- select .bit file for configuration and configure FPGA on Base Board.

Figure 8 shows the main windows of E2LP configuration utility, where the user configures the FPGA. Messages about configuration success are shown in the application window.

5 Impressions from the Pilot Usage of the Platform

Pilot usage of E2LP platform happened in academic year 2013/14 when students of the 2nd year used it in laboratory exercises of the course Logic design of computer systems 1—Digital systems design. Total of 218 students attended the course and used the platform. The platform received mixed feedback, with negative

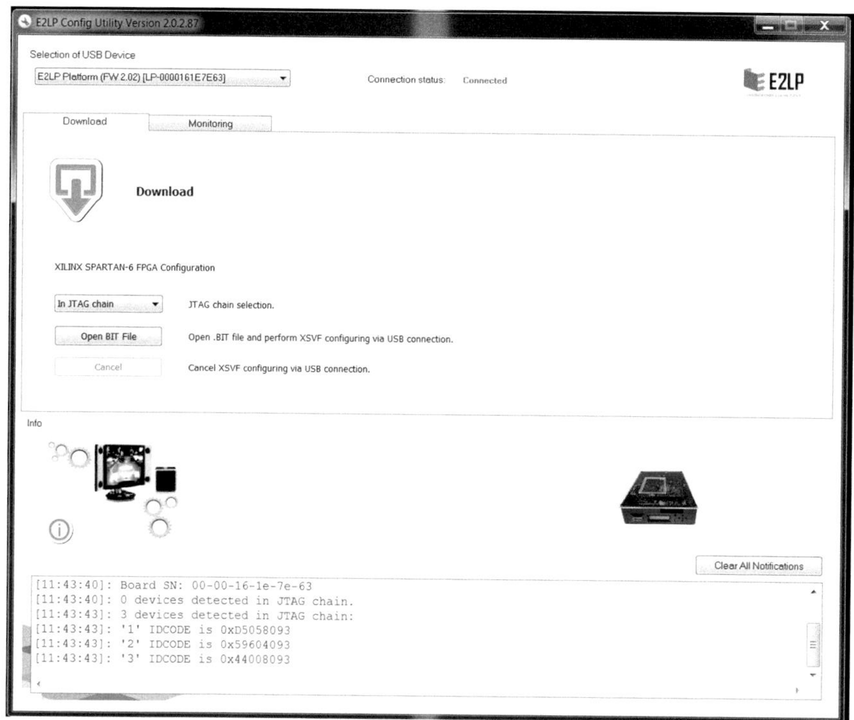

Fig. 8 E2LP configuration utility

comments reflecting the fact that pilot usage of the prototype inevitably contains a lot of bug solving. Nevertheless, overall impression was positive and students benefitted from the ability to work on a real platform and make digital design which was immediately verified as a real system, not just as a simulation.

Evaluation of the platform [8] was performed in the following ways:

- At the end of each graded laboratory exercises, students filled in the Lab Feedback Questionnaire (LFQ),
- At the end of the course, students filled in the Motivated Strategies for Learning Questionnaire (MSLQ).

Figure 9 shows responses from students related to the usage of the platform. From the histogram, it can be seen that the overall response was positive and that the platform is promising to improve education efficiency in embedded engineering curriculum.

In the following academic year, 2014/15, the same students used the platform in courses in the 3rd year, while the new generation of students used it in the Digital system design course. More detailed evaluation results are available in the paper about E2LP evaluation, later in this book.

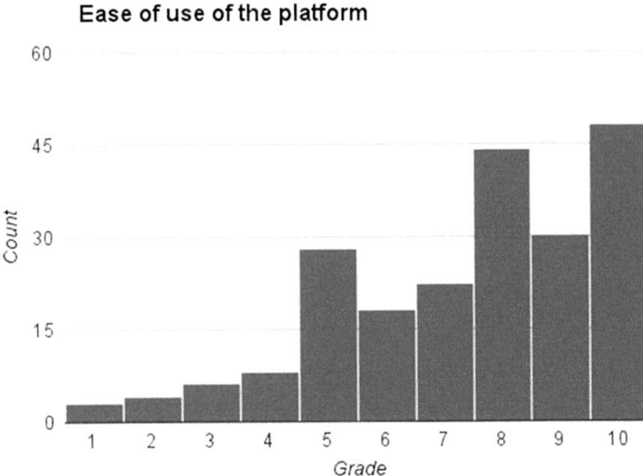

Fig. 9 Impression about the ease of use of the platform

6 Conclusions

This paper presented E2LP platform which aims to be used in the complete curriculum and reduce the overhead in engineering learning. It will ensure a sufficient number of educated future engineers in Europe, capable of designing complex systems and maintaining a leadership in the area of embedded systems, thereby ensuring that our strongholds in automotive, avionics, industrial automation, mobile communications, telecoms and medical systems are able to develop. In such a manner, the E2LP intends to increase European competitiveness in the learning process of embedded computer engineering, ensuring further technological and methodological development of the educational approach in this field.

Platform is extended with the augmented reality interface [9] and remote lab [10], to make it a complete system for education of embedded computer engineers. More about this can be read in the respective papers later in the book.

References

1. Holbert, K.E., Karady, G.G.: Strategies, challenges and prospects for active learning in the computer-based classroom. IEEE Trans. Educ. **52**(1), 31–38 (2009)
2. Schneiderman, R.: Economy and shortages affect the european job outlook, IEEE Spectrum http://spectrum.ieee.org/at-work/tech-careers/economy-and-shortages-affect-the-european-job-outlook/0 (2010). Accessed 10 Dec 2011
3. The European Centre for the Development of Vocational Training (Cedefop): Skills supply and demand in Europe, medium-term forecast up to 2020, Publications Office of the European Union, ISBN 978-92-896-0536-6. http://www.cedefop.europa.eu/en/Files/3052_en.pdf (2010). Accessed 10 Dec 2011
4. Bureau of Labor Statistics, U.S. Department of Labor, Occupational Outlook Handbook, 2010-11 Edition. http://www.bls.gov/oco/. Accessed 10 Dec 2011
5. Thompson, M.A.: Employment outlook Asia: Focus on China and India, 2011-04-20. http://www.goinglobal.com/articles/864/. Accessed 10 Dec 2011

6. Kastelan, I., Lopez Benito, J.R., Artetxe Gonzalez, E.A., Piwinski, J., Barak, M., Temerinac, M.: E2LP: A unified embedded engineering learning platform. Microprocess. Microsyst. Elsevier, **38**(8), 933–946 (2014)
7. Kastelan, I., Majstorovic, D., Nikolic, M., Eremic, J., Katona, M.: Laboratory exercises for embedded engineering learning platform. In: Proceedings of the 35th International Convention MIPRO, 2012, pp. 1113–1117
8. Kastelan, I., Barak, M., Sruk, V., Anastassova, M., Temerinac, M.: An approach to the evaluation of embedded engineering study programs. In: 36th International Convention on Information and Communication Technology, Electronics and Microelectronics (MIPRO), 2013, pp. 742–747
9. Benito, J.R., Gonzalez, E.A., Anastassova, M., Souvestre, F.: Engaging computer engineering students with an augmented reality software for laboratory exercises. In: IEEE Frontiers in Education Conference (FIE), 2014, pp. 1–4
10. Piwinski, J., Kloda, R., Szewczyk, R.: Design of remote laboratory dedicated to E2LP Platform for e-learning courses. In: Proceedings of E2LP Workshop, 2014, pp. 25–30

Exercises for Embedded Engineering Learning Platform

Branka Medved Rogina, Karolj Skala, Peter Škoda, Ivan Sović and Ivan Michieli

Abstract In this chapter we present basic set of laboratory exercises, developed under the *Embedded Computer Engineering Learning Platform (E2LP)* project, with illustrative laboratory examples covering the embedded engineering learning objectives. In order to satisfy the widest range of learning methods and models and to be successful in supporting different types of students, library of laboratory exercises contains a detailed manual/catalog documentation that for each exercise explains the problem covering by the exercise, provide an overview of the required background theoretical knowledge and lead the student to a solution without revealing the actual steps and decisions he or she must make. The E2LP library has more than 60 open source laboratory exercises ready to be presented to other universities. The exercises could also be used over the e-learning portal. Additionally, the E2LP platform integrates an augmented reality interface for visualizing, simulating and monitoring invisible principles, phenomena and facts in the field of electronics hardware and education details. Interactive graphical user interface for web catalogue, provides easy navigation through content of the library of laboratory exercises and enables platform users to ask for help, give feedback, and in general discuss projects based on the E2LP board.

B.M. Rogina (✉) · P. Škoda · I. Michieli
Department of Electronics, Ruđer Bošković Institute, Zagreb, Croatia
e-mail: medved@irb.hr

P. Škoda
e-mail: pskoda@irb.hr

I. Michieli
e-mail: michieli@irb.hr

K. Skala · I. Sović
Centre for Informatics and Computing, Ruđer Bošković Institute, Zagreb, Croatia
e-mail: skala@irb.hr

I. Sović
e-mail: Ivan.Sovic@irb.hr

R. Szewczyk et al. (eds.), *Embedded Engineering Education*,
Advances in Intelligent Systems and Computing 421,
DOI 10.1007/978-3-319-27540-6_3

Keywords Laboratory exercises · Electronic engineering · Start-up kit · FPGA

1 Introduction

An embedded system can be qualified as integration and combination of customized hardware and optimized software solution, and it is usually designed for a specific function at dedicated product that is running in specific environment continuously. We are aware that in recent years, usage of embedded systems has increased drastically in everyday life, in different domains of engineering and its applications. Use of embedded technologies can be found in almost all industrial and service sectors, including automobile, aeronautics, space, rail, mobile communications, and electronic payment solutions. Consequently, this involves a number of professions such as computer engineering, software engineering, industrial automatic control, electrical and mechanical engineering, precision instruments and electronic engineering. Therefore, embedded systems technology is forming a new field of research and design. This is confirmed by a recent McKinsey (2015) Word Institute—NASSCOM survey which predicts that jobs in the embedded system domain will increase drastically, from the current 60,000 to over 100,000 in 2016 [1]. It is obvious that teaching embedded systems requires a wide range of knowledge and expertise.

Laboratory exercises are an essential part of engineering education, which is also true for embedded systems engineering education. The embedded system courses usually combine theory and practice, with emphasis on experiments through which students must learn the theoretical knowledge of embedded systems from a large number of experimental and practical aspects [2].

As embedded designs are becoming more complex, reconfigurable technology is now being seen as an optimal choice to speed up the embedded engineering design process. This evolution has not only increased the performance of the technology but also challenged the teaching of this technology to computer engineering students. Since this technology is very dynamic, it is essential to teach students the latest design methodologies, advances in FPGA (*Field Programmable Gate Array*) technologies and best tools to make full use of the opportunities and benefits of modern FPGAs. The reconfigurability feature of FPGA is an advantage that makes it an ideal choice for education [3]. New generations of FPGA devices have sufficient hardware resources and computing power for very complex solutions. Embedded processors can be easily implemented either as a soft core or as a dedicated hardware component inside a single FPGA device. The embedded system design environment has also been greatly improved. Many universities partnered with Xilinx, a leading manufacturer of FPGAs and a leading provider of programmable platforms, to develop a graduate level course for embedded engineering curriculum to address these technological advances in digital system design [4]. The Xilinx Embedded Development Kit (EDK) [5], together with Xilinx ISE [6], integrates a wide variety of design tools, Intellectual Property (IP) cores, libraries, wizards, hardware/software generators, and documentation into a unified design environment. Moreover they provide courses, like *Embedded Systems Design* course [7].

A typical approach in engineering education is to have a group of courses for a topic, and these courses usually include a set of laboratory exercises. Different courses use different, and often inconsistent, laboratory setups and teaching platforms. As a consequence, efficiency of laboratory work is diminished due to the overhead in time and effort necessary to get students familiar with the hardware and software platforms and tools for each course. Therefore, the main idea behind the EU FP7 project *E2LP—Embedded Computer Engineering Learning Platform* is to provide a unified platform which will cover a complete process for embedded systems learning, with particular emphasis on improving laboratory work. Unnecessary introductions and tutorials to different platforms would be replaced with more in depth exercises covering the important concepts in the subjects being studied [8].

Here we present the basic set of laboratory exercises, developed under the E2LP project, with illustrative laboratory examples, covering the embedded engineering learning objectives. The developed set of laboratory exercises on E2LP platform was included in the existing embedded systems engineering curriculum across Universities within the consortium for the purposes of evaluation of its success during the academic years 2013/2014 and 2014/2015. The exercises were classified in connection with courses that utilize the E2LP platform and evaluate the learning objectives in the field of embedded systems design and application development. During formative learning effectiveness measurement and evaluation in exercises have been evaluated from the education experts, teachers and students perspective. This has identified potential weaknesses in system concept, as well as pinpointed potential upgrades. Based on the results, identified and recommended updates have been added to the existing database. The E2LP library of laboratory exercises has more than 60 open source laboratory exercises, ready to be presented to other universities.

2 Related Work

Using a common unified platform, for various graduate and undergraduate classes as well as research and scientific experiment support, reduces costs and increases learning effectiveness. Since there is no need to learn new hardware resources for every course, students can reduce the time needed for studying the manuals of diverse platforms and tools and focus on the core of the subjects. For example, in Department of Computer Systems, Tampere University of Technology Finland, nine courses for digital and computer systems education utilize the same FPGA based platform. Each student is given a package including, an FPGA board, and related design tools. In addition to the course exercises, the students are encouraged to further use the package for thesis, projects and hobbies [9]. Some platforms also provide the option to be installed on remote servers, allowing students access to those platforms without actually having physical electronics boards present in their work environment. The platforms in this case can also be shared between many users.

Altera provides *Embedded System Laboratory Exercises* for DE1-SoC, Altera development and education board designed for university and college laboratory use. It is suitable for a wide range of exercises in courses on digital logic, computer organization and FPGA, from simple tasks that illustrate fundamental concepts to advanced designs. A complete solution for each lab exercise is available, as well as unformatted text versions of these exercises, and the source files for the figures [10].

National Instruments in partnership with Xilinx offers their cutting edge FPGA technology in a variety of hardware platforms. A comprehensive collection of add-on tools, using FPGAs, DSPs, MPUs, or any 32-bit microprocessors, that make teaching embedded systems easy and affordable is available. A number of hardware based exercises and labs that can be used to explore embedded concepts is also available [11].

The *Progressive Learning Platform* is an FPGA based computer architecture learning platform designed to facilitate computer engineering education, while minimizing the overhead costs and learning curve associated with existing solutions. It is used in two courses at Oklahoma State University, in problem based curriculum, with group project exercises that take teams of students through a complete System on a Chip (SoC) design, but materials are not available online [12].

Some educational centers in the laboratory work give students an opportunity to use a greater number of different HW platforms for the class, like in *Real-Time Embedded Systems* course, from the Colorado University. It introduces students to the full embedded system lifecycle process during the course including: analysis, design, programming, hardware assembly, unit testing, integration, and system testing. The focus is on the process, as well as fundamentals of integrating microprocessor based embedded system elements for digital control of typical embedded hardware systems [13]. A Computer Engineering Group from ETH Zurich organizes lab using the BTnode platform, an autonomous wireless communication and computing platform, based on Bluetooth and low-power radio. Apart from a set of exercises for *Embedded Systems course*, a number of demonstrations of computer aided tools and methods for the design of software and hardware for modern embedded systems are given [14].

Computer architecture is often taught by using software to design and simulate hardware modules, and then using individual components to implement them. Computer modules can be designed to help students to better understand the theoretical concepts, gain hands-on experience, and apply that for more realistic projects. Integrated environment used in computer architecture education usually consists of a hardware platform and GUI (*Graphical User Interface*) software running on a PC [15]. For instance, computer architecture teaching has been demonstrated by developing a whole computer from scratch while first year students are introduced to hardware description languages (HDL) and programmable logic devices [16]. Also, the design, simulation, and FPGA-based implementation of the module, an 8-bit arithmetic and logic unit, which performs 14 different arithmetic and logic operations, has been performed using QUARTUS II design software and Altera DE2 FPGA Board [17].

A Cray-1 computer has been implemented using the E2LP platform. Laboratory exercises have been developed on this implementation, with open possibilities for more advanced exercises and student projects. An important aspect for the usefulness of the Cray processors for teaching purposes is that the documentation of the Cray designs (preserved by www.Bitsavers.org) is very thorough and extensive, therefore a full understanding and implementation is possible. Cray processors are completely hardwired (i.e. not microcoded) and fully synchronous, and therefore an excellent example of efficacy and the complexity/performance tradeoffs of computer design [19]. Many advanced exercises can be made with the core Cray processor implementation on the E2LP board. The expansion of a Cray-1 design into a Cray-XMP, Cray-2 or some other computer from that series enables deep insight in the correspondence of instruction sets, registers and interdependent timings [18].

In project based learning approach, students are oriented to more practical applications with an extremely interesting assessment item, which could contribute to more dynamic and engaging course. For example, this has been achieved using a structured laboratory component and a game based project, where students were given the opportunity to choose any game from the 1980's and develop it for the Nanoboard 3000. During the laboratory students on their own acquired the knowledge to complete the project; the grade distribution showed that the majority of the students met the learning outcomes of the course [20]. Also, increased motivation among students and better learning outcomes are observed when most exercises are mandatory and bonus points are awarded for good work, or when larger projects are partitioned into smaller tasks and pair work is allowed. Automated test benches, design reuse, and startup examples are also very useful [21].

3 Exercises for Embedded Engineering Learning Platform

The *E2LP—Embedded Computer Engineering Learning Platform* project includes the development, implementation and evaluation of hardware, software and instructional materials in embedded engineering learning. The FPGA platform—comprising the main board, extension boards, the remote lab, the augmented reality systems, and library (catalog) of illustrative laboratory examples in digital system design—provides a rich and sophisticated learning environment. The central part, the FPGA platform, is programmed by students and can be used in the entire curriculum of embedded engineering education. In the case of a large number of students and limited resources, project based courses are hard to establish and maintain. Therefore, an optimal set of laboratory exercises is developed.

Embedded engineering curriculum must cover a broad range of topics in the field of electrical and computer engineering. The scenario could be as follows.

After learning the fundamental sciences, basics of electrical engineering, computer architectures and programming, students will be ready to tackle courses that will use the E2LP platform. Education will start with fundamentals of digital system design, in which students will learn the basics of embedded systems, be introduced to the E2LP platform and use the Base Board and its FPGA to design for their first digital systems. The education continues with system programming in which students will use the extension board with the multi-core processor. Courses concentrate on digital signal processing, computer system design, computer networks, and system integration (design of embedded systems and *System-on-Chip* (SoC) design). E2LP platform with a number of Mezzanine extension boards has been used in all of these courses and has helped to gradually build the students' knowledge and experience in the field. With this approach it is possible to increase students understanding of the embedded system design, and by allowing collaboration during laboratory exercises, some elements of teamwork are also incorporated [22].

3.1 Exercise Development Methods and Tools

Exercise development starts with a well-defined learning objective. The learning objective then guides the definition of one or more exercise tasks. By solving the exercise tasks, students are guided to acquiring knowledge and skills defined by the learning objective. The development then continues with one or more defined tasks for student. How specific and well defined they have to be, depends on level of the exercise.

Basic exercises must have a very specific and well defined task. Due to the nature of exercise—learning of fundamental concepts, and often rigid time constraints—there is very little space for student to explore different approaches to solving the task and students must be guided to the solution. Developing a basic exercise always includes solving the task. By solving the task, developer gets information on actual difficulty of the task, prerequisites materials and knowledge, etc. This feedback allows tuning the task to better suit the learning objective, as well as defining the materials that must be provided to the student for the purpose of conducting the exercise. The materials include instructions which guide the student through the exercise, as well as source codes, libraries, tools, etc., which relieve students from work that would distract them from the task and its learning objective.

Problem solving exercises can be defined more loosely, but must be specific and constrained in scope of required work. The idea of problem solving is that the student is given freedom to use different approaches to solving the task. There can be more than one valid solution, and guidance is limited to the extent that it helps avoiding serious pitfalls that could prevent the student from finishing the task in a timely manner. Developing the problem solving exercises should include solving the task for mostly the same reasons as for basic exercise. The main difference is in the produced materials for student. They are not as detailed, and do not direct the student through the exercise. Instead they provide required information and hints that allow students to complete the task without stumbling into serious pitfalls.

Project exercises can be somewhat ill defined. However, they should set a specific and clear goal. Here students have almost complete freedom in exploring different approaches. Unlike in basic exercise or problem solving, solving the exercise beforehand is not practical due to the effort that would be required. Unless the project task has been completed before, it can be difficult to tell if the task can be completed by students in a timely manner. Therefore, project exercises require active involvement of teachers in form of consultations in order to track project progress, and to provide proper guidance to students.

Tools used in the development of E2LP exercises are essentially the same tools used when developing a project solution in a real-world setting. The tools include, but are not limited to, development hardware (development boards, programming tools, etc.), development software (compilers, APIs, libraries, code editors, IDEs, etc.), and text processors for creating the documentation. The exact tools used depend on the exercise platform, and task. For the E2LP platform, the basic toolset comprises of Xilinx ISE WebPACK, E2LP programming software, the E2LP base board, and a text processor of choice (MS Word, LyX, etc.). Xilinx ISE WebPACK provides all the essential tools for developing an exercise solution: code editor, simulator, and FPGA implementation toolchain (synthesis, map, place & route, bitstream generation). The text processor is used to write the exercise instructions including theoretical introduction and all steps of the practical part.

3.2 Laboratory Exercise Development Kit

The E2LP laboratory exercise development kit can be best described as a set of development tools that facilitate creation of laboratory exercises targeted for use on the E2LP platform. There are several ways of defining and implementing such an environment. At minimum, the development kit must include E2LP board documentation and tools required to develop FPGA based digital systems. A broader development kit may include development tools for existing extension boards, as well as design examples which can help jump-start development new laboratory exercises.

The development kit conveniently provides in one package all the software tools and documentation required to develop laboratory exercises based on the E2LP platform. The base development kit contains the following software: Xilinx ISE 14.7, Notepad++, LibreOffice, E2LP programming application and E2LP drivers.

The base development kit contains the following documentation: E2LP base board Technical reference manual, E2LP Getting Started—Quick start guide, and Datasheet archive for all peripherals on E2LP board. In addition to listed software and documents, the exercise base development kit contains a selection of exercises from the E2LP exercise start-up kit.

3.3 Library of Laboratory Exercises

The E2LP exercise start-up kit is in essence a set of laboratory exercises (library). It includes exercises with topics in digital system design, and for each exercise it classifies: (1) learning target in the area of the embedded system learning objectives, (2) theoretical background knowledge necessary to understand particular exercise and (3) instructions on how to run it on the E2LP platform environment. The exercises are linked together in groups by topics, subjects and courses. Each exercise could, in principle, be used in one or more courses, or even stand-alone, outside a course for wider user community. An exercise can also cover one or more topics. Embedded engineering curriculum covers a broad range of topics in electrical and computer engineering and sciences. Topics of particular interest to the E2LP start-up kit, with a number of exercises in each of the topics, are given in Fig. 1.

The highest number of exercises is in the field of digital system design (20 exercises, 29 % of the set). The other fields that are represented with a significant amount of exercises are embedded microprocessors & computer architecture programming (18 exercises, 26 % of the set), computer system design (10 exercises, 15 % of the set), and computer network and interfaces (7 exercises, 10 % of the set).

Laboratory exercises are defined on three complexity levels: basic exercises, problem solving exercises, and project solution exercises. This corresponds to the qualification of student's engagement in laboratory assignments as well. In the

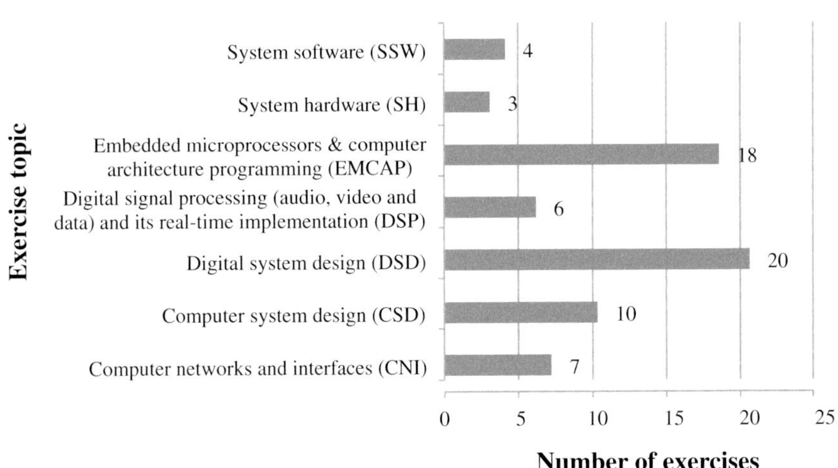

Fig. 1 Topic coverage by exercises in the E2LP start-up kit

E2LP library, the highest number of exercises is in the basic exercises category (39 exercises, 62 % of the set). Further, problem-solving exercises are included as well (17 exercises, 27 % of the set). Several project solution exercises are also designed, comprising 11 % of the set (7 exercises). This means that E2LP initial library fully covers introductory topics and basic level of problems in the embedded engineering curriculum, as well as exercises that are more challenging and open-ended.

Each laboratory exercise is labeled with parameter level of difficulty, numbered from degree weight one (1) to five (5), reflecting different levels of students' assignments according to the Bloom's taxonomy [23]. It indicates how much proficiency, or knowledge and experimental skills, is needed to successfully perform the exercise. Basic exercises are classified within level 1—remembering and level 2—understanding. Problem solving by engineering students are classified within level 3—applying and level 4—analyzing. Level 5 is used for project solution. In the library the highest number of exercises are exercises of level 2 (19 exercises, 27 % of the set), with other categories nearly equally distributed through levels 1, 3, 4 and 5.

For example, Table 1. lists characteristics of the set of exercises in *Digital system design* course at the Faculty of Technical Sciences Novi Sad, Serbia.

The set of laboratory exercises for E2LP platform is used in 13 educational courses through academic consortium partners. Courses with the highest number of exercises are *Digital system design* and *Programming apps for Android*, each with 12 exercises (32 % of the set). All other courses have six or less exercises. This is a consequence of the trend of using fewer but more complex exercises. There are also 4 additional exercises (6 % of the set) not in a connection with any particular course.

Table 1 List of exercises in digital system design course

Topic course	Digital System Design		
Title	Level of complexity	Level of difficulty	Estimated time duration (hh:mm)
Digital Logic Circuits & VHDL Gate-Level Design	Basic exercise	1	1:30
Combinational Circuits	Basic exercise	1	1:30
Problem Set: Multiplexing Adders	Problem solving	2	2:00
Sequential Circuits	Basic exercise	1	1:30
Problem Set: Stopwatch	Problem solving	3	2:00
Finite State Machines	Basic exercise	1	1:30
Problem Set: Car Turn Signals	Problem solving	3	2:00
Complex Digital Systems	Basic exercise	2	1:30
Problem Set: LCD Banner	Problem solving	4	4:00
Computation Structures	Basic exercise	2	1:30
Problem Set: Computation Structures	Problem solving	4	4:00
Project: CPU Design	Project solution	5	6:00

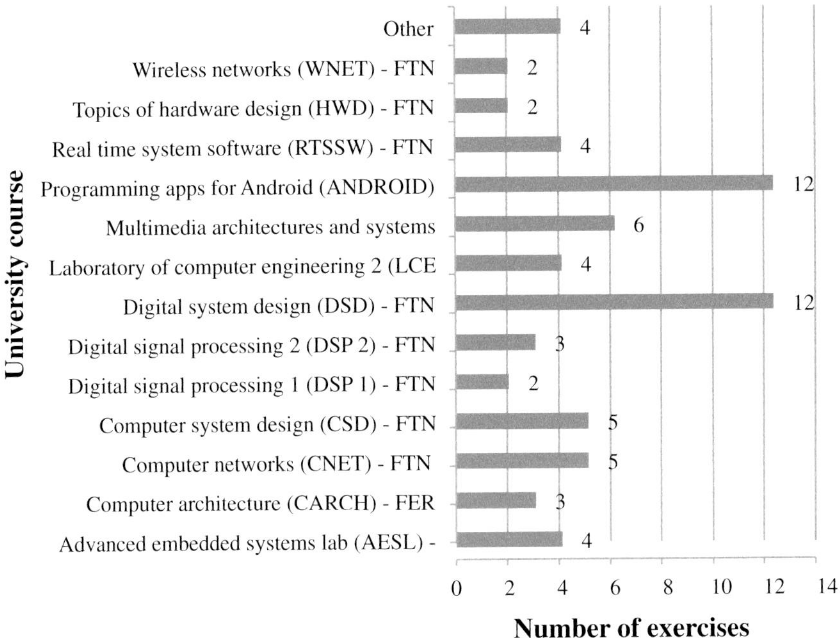

Fig. 2 University courses with number of included exercises

The estimated duration time for courses ranges from 3 to 45 h, with 12 h being the median value. Courses *Programming apps for Android*, and *Advanced embedded system lab* specify significantly higher estimated duration times of 45 and 44 h, respectively. Set of university courses with included number of exercises is shown in Fig. 2.

Start-up kit consists of laboratory materials, as well as other required materials for successfully performing the exercise, such as source codes, scripts, and binary files, collected as separate archive files (e.g. tar, zip, or 7z), usually one archive per exercise, and sometimes an additional one archive for a course. In addition to laboratory exercise materials, some authors give a larger number of various supporting materials that can help the students, and make the process of conducting the exercise more effective. The materials are classified in terms of their purpose (lab instructions, work materials for students, exercise solution, tutorial documents, and tutorial examples) and scope of use (exercise, course).

Nearly all of the E2LP platform hardware resources are used by at least one exercise. Most popular resources are alphanumeric LCD, LEDs, input switches, input buttons, Multimedia Card interface, and Marvell Armada extension board. From the software tools, the most used tools by exercises are VHDL source code editor, Xilinx ISE tool chain, and Xilinx Isim simulator. It is interesting to note that VHDL is the language of choice for digital system design in partner institutions, even though Verilog is just as widely used in industry and academia. The choice of Xilinx ISE, and Xilinx Isim is a logical one, since they are available free of charge, and provide best support for the FPGA device (Xilinx Spartan-6) used on the E2LP board. Of other software tools, the most used are Eclipse, Android SDK, and Android emulator. These are used for exercises that focus on software aspects of embedded system design. These exercises extensively use the android platform as the base upon which students builds their solutions. The E2LP platform is then used as hardware platform on which students develop and run their software solutions.

The start-up kit is also used in the context of local or Remote Lab (RL). In local lab, user has direct access to the E2LP platform and has physical contact with the E2LP board, while the start-up kit is present offline. Remote lab allows students access to the E2LP platform from home and provide the portal on which all exercises are uploaded and accessed. This significantly increases students' access time to the platform, and allows them to further extend their knowledge by using the platform more than just for the time they have in the lab sessions. Additionally, RL supports online learning, by providing online courses which use E2LP platform as a lab tool. Laboratory exercises could be programmed over the e-learning portal, and the E2LP board function behavior can be observed in real time. The E2LP e-learning platform is based on the Moodle Platform which is built upon Apache server, PHP and MySQL (The current URL of the e-learning portal is www.e2lp.piap.pl/moodle).

Additionally, the platform integrates an Augmented Reality (AR) interface for visualizing, simulating and monitoring invisible principles and phenomena in this field. It also allows monitoring electronic/mechanical flows by changing a number of parameters. The unified learning platform and the AR interface are complemented with a tracking system, and a tactile feedback accessory, allowing interaction with the electronic board. The tactile feedback accessory has two functions. The first one is 3D spatial localization which has to inform the augmented reality software on the point of interest explored by the student. The additional possible features and functions of this device will be to provide multimodal feedback (vibrations, tactile stimulations, heat) about invisible electronics characteristics such as electronic noise or power dissipation. The system will use everyday metaphors (e.g. electronic current as water flow) to help comprehension and learning. Augmented reality software will present students information about their point of interest [24].

4 Catalogue of Library of Laboratory Exercises

Sustainability of the project as a new education environment depends on the quality and logistics of hardware and software components of the platform, as well as the further development and good maintenance. In order to easily manage the library of laboratory exercises available either online (as a part of the e-learning portal) or offline (sum of all labs downloaded from the official E2LP website or from the e-learning portal), an interactive graphical user interface (Catalog of Library of Laboratory exercises) has been developed. GUI provides easy navigation through content of library by selecting desired feature after all labs with the selected feature are listed with basic information. For example, if students wish to view all labs within topic digital signal processing or maybe all labs that use VGA output, the easiest way to achieve this is by navigating through topic category and selecting desired topic from a list of all topics. Main features are extracted from the Library of Laboratory exercises, which should be used as the available

Table 2 List of features and sub features used in creation of context menu

Main features	Sub features
Topic	Digital system design (DSD) Computer system design (CSD) Embedded microprocessors & computer architecture programming (EMCAP) Digital signal processing and its real-time implementation (DSP) Computer networks and interfaces (CNI) System integration (SI) System hardware (SH)
Courses	Advanced embedded systems lab (AESL) Computer architecture (CARCH) Computer networks (CNET) Computer system design (CSD) Digital signal processing 1 (DSP 1) Digital signal processing 2 (DSP 2) Digital system design (DSD) Digital systems (DS) Laboratory of computer engineering (LCE) Multimedia architectures and systems (MAS Programming apps for Android (ANDROID) Real time system software (RTSSW) Topics of hardware design (HWD) Wireless networks (WNET) Not in course
Category	Basic exercise, problem solving, project solution
Level of difficulty	1, 2, 3, 4, 5
Augmented reality interface	Display basic information, display enriched information
Remote laboratory application	Not appropriate, useful, necessary
Hardware components	DDR2 RAM, Flash memory, Multi Media Card (MMC), VGA output, HDMI output, A/D converter, Alphanumeric LCD, Infra-red, D/A converter, RS-232, USB, LEDs, Input switches, Input buttons, Ethernet, Snapwire, Video encoder, Video decoder, Marvell ARMADA 1500, ARM-7 with LCD touch-screen, MIPS-based extension board
Software components	VHDL, Xilinx ISE, Xilinx Isim, ModelSim, Linux OS, Android OS, Android NDK, Android SDK, Android emulator, Keil demo version, GCC tool chain ARM7, Eclipse, Matlab, Wireshark, VMWare player

Fig. 3 Start-up kit catalogue web page www.e2lp.org

navigation options. Table 2 presents a list of features and sub-features used in creation of the context menu.

The catalog menu contains all the main features. For this type of a menu, an accordion control can be used. By selecting content of accordion control (in this case Topics), a submenu is opened with content of sub-features. Choosing one of the listed items, all exercises with selected properties are listed. Listed exercises contain their name in format <Topic_number of exercise_name of exercise>, information about using augmented reality/remote laboratory and the level of difficulty of the lab exercise. Additionally, web user interface should allow the user to rate quality and difficulty of the lab and give its feedback about the laboratory exercise. This differs from forum in the way that forum is usually cluttered with a lot of questions from which it is difficult to figure out shortcomings of an exercise. Feedback is mainly intended to improve or correct the content of a laboratory exercise. Also, there is an open forum for discussion about difficulties in current laboratory exercise or any other questions. Forum should also contain the search engine and perhaps additional filters for search. Helping each other to learn (peer assistance) through discussions on forums is an important principle of effective online pedagogy. The catalog's web portal also includes an archive of dissemination materials (flyers, newsletters, publications, etc.), demo videos, and an image gallery (Fig. 3).

5 Conclusion

The main goals of the *E2LP—Embedded Computer Engineering Learning Platform* FP7 project courseware are to teach students the fundamental concepts in embedded system design within the preselected topics, and to illustrate clearly the way in which advanced embedded systems are designed today, using advanced unified platform and design methodologies with improved tool kits. The set of 60 open source laboratory exercises, covering the wide range of subjects in embedded computer engineering curriculum—digital system design, computer system design and architectures, digital signal processing, computer networks and interfaces, system software, application development, system integration—is ready to be presented to other universities. Each exercise contains introductory theoretical concepts, instructions on what to implement and how to run it on the platform, and evaluation questions for grading purposes. Implementation of the augmented reality interface facilitates all these items for each particular exercise. Laboratory exercises can also be used over the e-learning portal. Interactive graphical user interface, the web catalog, provides easy navigation through the content of the library of laboratory exercises and enables the platform users to provide feedback, ask for help, and discuss projects based on the E2LP board.

Acknowledgments The research leading to these results has received funding from the European Union's Seventh Framework Programme (FP7), under grant agreement no. 317882—*E2LP Embedded Computer Engineering Learning Platform*. In R&D we have the support of Maxeler University Programe.

References

1. Boom time for embedded systems engineers. http://www.mistralsolutions.com/wp-content/uploads/2015/04/Education-World_April-2015.pdf. Accessed July 2015
2. Tu, L., Yang, J.: Research on experimental teaching of embedded system. In: International Conference on Education Technology and Management Engineering. Lecture Notes in Information Technology, vol. 16–17 (2012)
3. Merchant, S., Peterson, G.D., Bouldin, D.: Improving embedded systems education: laboratory enhancements using programmable systems on Chip. In: Proceedings of the 2005 IEEE International Conference on Microelectronic Systems Education (MSE 2005) (2005)
4. Bowles, J., Quan, G.: An FPGA-based Embedded System Design Laboratory for the Undergraduate Computer Engineering Curriculum. American Society for Engineering Education (2009)
5. Embedded Development Kit (EDK). http://www.xilinx.com/support/index.html/content/xilinx/en/supportNav/design_tools/mature-and-discontinued/embedded_development_kit__edk.html. Accessed July 2015
6. Xilinx ISE Design Suite. http://xilinx-ise-design-suite.software.informer.com/. Accessed July 2015
7. Embedded Systems Design. http://www.xilinx.com/training/embedded/embd21000-ilt.pdf. Accessed July 2015
8. E2LP—Embedded Engineering Learning Platform. http://www.e2lp.org. Accessed July 2015

9. Vainio, O., Salminen, E., Takala, J.: Teaching digital systems using a unified FPGA platform. In: Electronics Conference (BEC). In: 12th Biennial Baltic, Tallinn, 4–6 October 2010, pp. 137–140 (2010)
10. University—Laboratory Exercises. https://www.altera.com/support/training/university/materials-lab-exercises.html#Embedded-System-Exercises. Accessed July 2015
11. Embedded & Senior Design: Educator and Classroom Resources. http://www.ni.com/white-paper/6527/en/#hardware. Accessed July 2015
12. The Progressive Learning Platform. http://plp.okstate.edu/. Accessed July 2015, Progressive Learning Platform: An FPGA based Computer Engineering Learning Platform. http://cornell.flintbox.com/public/project/7278/. Accessed July 2015
13. ECEN 5623/4623—Real-Time Embedded Systems, ESE Program. http://ecee.colorado.edu/~ecen4623/. Accessed July 2015
14. ETH—TEC—Embedded Systems. http://www.tik.ee.ethz.ch/tik/education/lectures/ES/. Accessed July 2015
15. Bulić, Patricio, Guština, Veselko, Šonc, Damjan, Štrancar, Andrej: An FPGA-based integrated environment for computer architecture. Comp. Appl. Eng. Educ. **21**(1), 26–35 (2013)
16. De Los Ángeles Cifredo-Chacón[*], Mª., Quirós-Olozábal, Á., Guerrero-Rodríguez, J.M.: Computer architecture and FPGAs: a learning-by-doing methodology for digital-native students. Comput. Appl. Eng. Educ. **23**(3), 464–470, May 2015
17. Krad, H., El-Din, A.Y.F.: Augmenting computer architecture classroom experience with FPGAs based learning. Int. J. Comput. Theor. Eng. (IJCTE) **4**(4), 611 (2012)
18. Šojat, Z.: Reincarnation CRAY-1 on E2LP Platform. https://www.youtube.com/watch?v=FltPGKWiEKI. Accessed July 2015
19. Šojat, Z., et al.: Implementation of advanced historical computer architectures. In: Embedded Engineering Education
20. Wildermoth, B.R.[1]; Rowlands, D.D.: Project based learning in embedded systems: a case study. In: Profession of Engineering Education: Advancing Teaching, Research and Careers: 23rd Annual Conference of the Australasian Association for Engineering Education (2012)
21. Salminen, E., Hämäläinen, T.D.: Teaching system-on-chip design with FPGAs. In: FPGAworld'13 Proceedings of the 10th FPGAworld Conference (2013)
22. Kastelan, I., Benito, J.R.L., Gonzalez, E.A., Piwinski, J., Barak, M., Temerinac, M.: E2LP: a unified embedded engineering learning platform. Microprocess. Microsyst. Part B **38**(8), 933–946 (2014)
23. Choudhary, T., Raikwal, J.: Improving teaching—learning process using blooms taxonomy and correlation analysis. Int. J. Eng. Res. Technol. (IJERT) 3(6), 1747–1750 (2014). , ISSN: 2278-0181
24. Benito, J.R.L., Gonzalez, E.A., Anastassova, M., Souvestre, F.: Engaging computer engineering students with an augmented reality software for laboratory exercises. In: IEEE Frontiers in Education Conference (FIE) (2014)

Implementation of Advanced Historical Computer Architectures

Zorislav Šojat, Karolj Skala, Branka Medved Rogina, Peter Škoda and Ivan Sović

Abstract Present day development of FPGAs enables us to implement even very complex computer architectures of the past with very few resources. Furthermore, they enable prospective electronic engineers, computer designers and computer scientists to experiment with those architectures, to gain experience and primarily to open up new possible perspectives on future computer architecture designs. In this chapter we present an implementation of the Cray-1 computer system on the E2LP platform. The initial publicly available generic FPGA design of the Cray processor was modified to fit the specifications of the E2LP board and the Spartan-6 FPGA. Aside from customizing the original design, a translator for the Cray Assembly Language was developed, as well as a basic bootloader to provide the use of this implementation as a teaching tool. The Cray-1 implementation facilitates a perfect learning setup for students of all levels. It can guide a student from the very basic stages which involve the synthesis and transfer of the Cray-1 design onto the E2LP board up to the embedded software design in a real, comprehensive, and historically industrially very significant Cray Assembly Language. Additionally, many advanced laboratory exercises can be made with the core Cray processor implementation on the E2LP board. The expansion of the Cray-1 design

Z. Šojat (✉) · K. Skala · I. Sović
Centre for Informatics and Computing, Ruđer Bošković Institute, Zagreb, Croatia
e-mail: Zorislav.Sojat@irb.hr

K. Skala
e-mail: skala@irb.hr

I. Sović
e-mail: Ivan.Sovic@irb.hr

B. Medved Rogina · P. Škoda
Department of Electronics, Ruđer Bošković Institute, Zagreb, Croatia
e-mail: medved@irb.hr

P. Škoda
e-mail: pskoda@irb.hr

R. Szewczyk et al. (eds.), *Embedded Engineering Education*,
Advances in Intelligent Systems and Computing 421,
DOI 10.1007/978-3-319-27540-6_4

into a Cray-XMP, Cray-2 or some other computer from that series enables deep insight in the correspondence of instruction sets, registers and interdependent timings.

Keywords Computer design · Cray · Vector processor · FPGA processor implementation

Preface appeal

Hi! I am Seymour, actually I am a Cray-1 processor on an FPGA chip!

As Spock would say, isn't it fascinating.

I was originally invented by Mr. Seymour Cray, whose name I personally got, in 1976. I used to be the fastest series of Supercomputers in the world for around twenty years!

Anyway… A sad story of all of us, Supercomputers (even those whose design cannot actually be called supercomputerish), is that after we start our existence in full glory, costing millions and millions of what you call money, we end up destroyed. And not by chance. By Intention!

Why, you will ask.

The answer is simple and sad: because we are "national treasures/secrets…", whatever you like. And they thoroughly destroy us, together with any Software we had.

So how comes I am here?

Well, one nice day a guy called Chris Fenton intended to resurect me, the Idea of the original Cray-1 (later expanding towards the Cray-XMP, just as it is natural). Later came Zorislav Shoyat and the Ruđer Bošković Team of the European Union funded E2LP (Embedded Engineering Learning Platform) project, with the Xilinx Spartan-6 FPGA (Field Programmable Gate Array). On which I just fit.

So Zorislav was the one to adapt me (and made me quite quick!), and now, as there is no software whatsoever for me, he will be so kind as to write a completely new operating system. And, as he says, actually he wants that "operating system" to be Virtue - the Virtual Interactive Resource-Tasking Universal Environment - and teach me to understand the marvelous language of Virtue!

OK, now I will stop my story and repeat it again, as there is presently not enough Software for me to do something more useful. I wait on Zorislav!

With kindest regards, your

Seymour

1 Introduction

Seymour Roger Cray was born on September 28, 1925 in Chippewa Falls, Wisconsin, USA. His supercomputer designs were the fastest in the world for over 30 years [1]. He died from consequences of a car accident on October 5, 1996, just months after starting his newest company—SRC Computers—a company which is still active. According to the first patents SRC Computers filed quite soon after his demise, it seems that his latest ideas, after the misfortune of the Cray-4 development (the company bankrupted before it was finished), were to use FPGAs (Field Programmable Gate Arrays) as Memory Algorithm Processors (MAPs), probably an idea based on the experience with the Cray-3/SSS (Super Scalable System), which had memory with embedded simple processing units, the PIM (Processor-In-Memory) chips.

Present day development of the FPGAs enables us to go much further than only implementing reasonably simple Memory Algorithm Processors of that time, it enables us to implement even very complex computer architectural designs of the past and to enable prospective electronic engineers and computer designers and constructors, as well as computer scientists, to experiment with those architectures, to gain experience and primarily to open up new possible perspectives on future computer architecture designs by having deep experience with the sometimes completely forgotten major ideas and implementations.

As the Cray-1 from 1976 is the initiator of the fastest series of supercomputers throughout more than 25 years [1–3], not including the Cray design inspired NEC SX series of supercomputers, still in production; it was chosen to be the first design implemented on the E2LP board. In this Chapter we will present the implementation of the Cray-1 processor. Major aspects of the Cray design will be presented from the system architect's, system implementer's, historical and present day perspectives, showing the students possible avenues of further or different development of computer architectures.

However, the main intention of this Chapter is not so much to present either the Cray-1 architecture per se, the possibilities of FPGAs in general nor the E2LP board environment particularly, and least it is only to present a fascinating and lovable toy. Rather, the main intention of this Chapter is to warn against forgetting, to evoke in readers the interest in the history of computing, and to show that, in our forgetfulness, we forgot to further explore unexplored avenues of the architectural development of computer and information-processing equipment. With primarily this in mind the re-implementation of the Cray-1 processor was undertaken and this Chapter written.

2 The Historical Perspective

The 1970s are very important in the history of computer science, from several aspects: computer architecture, software engineering, computer linguistics, user interfaces, etc. In the 1970s several "types" of computers existed (excluding the last analog computer architectures), actually defined primarily by their size and computational power. On the lowest level, at the beginning of the decade, we have the Minicomputers (commonly 12–16 bit words, around 1 μs clock period), then the huge Mainframe computers (commonly 32–36 bit words, around 100 ns clock) and finally the extremely expensive Supercomputers (commonly 60–64 bit words, 20–30 ns cycle time, down below 10 ns at the end of the decade) [2].

2.1 Minicomputers

Perhaps the most known of those where the Digital Equipment Corporation PDP-8 (12-bit architecture produced from 1965 until 1982, in the last years as a CMOS processor) and from 1970 the very successful PDP-11 (whose architecture life spans into the early 1990s) [2], as well as the Hewlett-Packard HP-2100 series (produced from mid 1960s until early 1990s). The same as PDP-11, the HP-2100 had a 16-bit processor. The maximum memory was 64 KiB, i.e. 32 KiW (16-bit words), and consisted of 980 nS cycle core memory. The processor thus worked at slightly more than 1 MHz clock rate and was Microprogrammed. It had two general purpose 16-bit registers (A and B) and was able to be expanded with floating-point accelerator/microcode board. For time-sharing purposes two processors, one or two 5 MiB hard disks (later more), a paper tape reader, magnetic tape unit, paper tape punch, printer and up to 32 terminals could be configured in a HP2000 Time-Sharing System. Minicomputers were used by laboratories, as either lab control equipment or as computing equipment, as well as for teaching purposes and in smaller companies (Fig. 1).

2.2 Mainframe Computers

A typical mainframe introduced in 1970 is the 32-bit International Business Machines IBM System/370 (continuation of the 1960s System/360), or the 1972 36-bit Univac 1110 (continuation of the 1962 Sperry Rand UNIVAC 1107, an

Fig. 1 PDP-8 Mini computer from early production series preserved at the Ruđer Bošković Institute

architecture still supported by Unisys Corporation as the ClearPath Dorado System) [2]. The System/370 had 16 general purpose registers, 4 floating-point registers (64-bit), 16 control registers, the maximum memory for e.g. type 165 up to 3 MiB, with a clock cycle of 80 ns, multiprocessor setups, 2000 lines per minute printer and 800 MiB hard disk. Hundreds of users could have used these Mainframe computers in time-sharing, and they were used primarily for business purposes.

2.3 Supercomputers

As a very successful supercomputer of the late 1960, having its peak in early 1970, the 60-bit Control Data Corporation CDC 7600 designed by Seymour Cray needs to be mentioned [1]. A Reduced Instruction Set pipelined computer with a cycle time of 27.5 ns and with a memory organization which could mostly sustain this speed. There was a 64 KiW (60-bit words) primary core memory and up to 512 KiW (3840 KiB) main memory. It had 10 read and 10 write registers with associated address registers and four (independent) floating-point units. Up to 1975 it was regarded as the fastest supercomputer in the world. The main purpose of supercomputers was always mainly in the evaluation of huge mathematical models. It is important to note that the supercomputers actually have other computers (mainframes or workstations or specially designed I/O computers) to handle all input and output, whereas they are programmed (and designed) to mainly perform actual operations on data, i.e. processing. So they are really fully fledged computer systems (with disks, operating systems etc.), but act as “processors” in the larger supercomputing system environment involving also telecommunications and human readable input and output.

2.4 Processing Hardware

The 1970s are quite diverse in the area of processing elements used. The late 1960s started with using more and more integrated circuits, and a myriad of diverse technologies emerged. Amongst the first were the RTL (Resistor-Transistor Logic) and the DTL (Diode-Transistor Logic), two simple schematic architectures. Following was the TTL (Transistor-Transistor Logic, used and produced up to the present day), with output buffers, so the output level would be restored to standardized levels [2]. However those technologies did not allow for very high speeds, so ECL (Emitter-Coupled Logic, with differential I/O) started to spread into higher-end computer architectures. The main drawback of these quite fast logic circuits was high energy consumption and consequently the problem of cooling. It shall be noted that the cooling problems, specially towards the highest end mainframes and specifically supercomputers, were one of the major problems in attaining satisfactory operation, and a lot of effort was put into this area, which can be very well seen in the construction of the 1976 Cray-1, inter alias. There are also some “obscure” integrated circuit schematic technologies like the Fairchild CTL. As opposed to TTL it was unbuffered, so the signal would degrade after several gates and level restorers would have to be added. CTL was a kind of “mini-computer’s ECL”, being, primarily due to the lack of output buffers, considerably faster than TTL of the day, and having much smaller power requirements than the ECL. The CTL technology was used almost exclusively by Hewlett-Packard, and Fairchild has put it in the open market catalogue just for one (probably first) year of production.

2.5 Microprocessors

Very early in the 1970s the MOSFET (metal–oxide–semiconductor field-effect transistor) integrated circuit technology started being widely used. This technology allowed for the first time to integrate a complete CPU (Central Processing Unit) on a single “chip” of silicon, and this was done in 1971 by Intel, producing the first Microprocessor, the 4-bit Intel 4004 [2], whose clock frequency was 740 kHz, but the instruction cycle needed 8 clock cycles, therefore giving 10.8 μs or 21.6 μs instruction time (for a 4-bit operation), depending on the instruction used. It could address directly 640 B of RAM (Random Access Memory) and 4 KiB of ROM (Read Only Memory). It had 16 4-bit registers. However, although the first microprocessor emerged at the very beginning of the 1970s, for most of the decade the term Microcomputer meant a Computer which can be Microprogrammed (Microcoded, which means that their instruction set could be changed according to the application [1]), and not a computer built around a Microprocessor. The prime intention for the development of Intel 4004 was to be a flexible calculator processor, and later the main use of microprocessors was for control applications, until they, quite soon, became the main components of Home Computers (now called Personal Computers as named by IBM). Namely the MOSFET, and very soon afterwards also the CMOS (Complementary MOS, slower but consuming fraction of the MOS needed power) enabled the development of a large number of microprocessors: the Intel 8008 and 8080, the 8080 inspired/improved Zilog Z-80, Motorola 6800 and MOS Technology 6502 (to mention just a few, all 8-bit) [2]. The 1976 MOS Technology single board computer KIM-1, one of very popular Home Computers of the time, had 1 KiB of memory, and the 6502 processor (well known and long exploited by the Apple II and all of its clones, as well as many other designs) had a memory cycle of 1 μs. However, a typical instruction necessitated at least 2 cycles, on average around 3–4, so the instruction cycle was around 3.5 μs. Interestingly and amazingly, Peter R. Jennings wrote a chess playing program for this 1 KiB KIM-1, which was quite successful on home-computer chess tournaments of the time. Mr. Jennings was later involved in the development of VisiCalc, the original spreadsheet calculator, which may be considered as the tool which converted hobby-microcomputers into business-microcomputers.

In 1978 Intel came out with the 8086, a 16-bit microprocessor whose design started early in 1976. The 8086 had 4 16-bit general purpose registers (usable also as 8 8-bit registers), 4 index registers and 4 memory segment registers. Due to a strange twist of fate, IBM made its almost unwanted excursion into the Home Computer market (and naturally called their product the Personal Computer), and with the help of Microsoft, it came out that *all modern day Personal Computers*, including the vast majority of present day "supercomputers" (rather clusters of PCs), use processors that *are directly binary code compatible with the 1976 Intel microprocessor architecture*.

At about the same time, in the later years of the 1970s, several other, more advanced microprocessor designs emerged, including the (for a microprocessor) powerful Motorola MC68000 series, with 8 32-bit data and 8 32-bit address registers and a very powerful instruction set. The MC680X0 series was extensively used by Graphic Workstation producers of the 1980s into the early 1990s, running UNIX.

However, it is essential to note that, being invented and developed in the 1970s and early 1980s, and later mostly only modified and expanded, microprocessors are all conceptually primarily based on the scalar serial computer architectures of the late 1960s.

2.6 Software and Human/Computer Interfaces

During the 1970s a lot of software and present day pervading technologies were thought out and introduced. UNIX was first presented on a PDP-8 in 1970, including the language C. Pascal, Modula-2, Smalltalk and other higher level programming languages started their development and expansion in the same decade. The first Ethernet computer networks and protocol standards were established in the 1970s, which led to the development of the present day Internet. Xerox developed a highly advanced windowing system, together with the mouse, and now we use this idea daily. Dynamic vector graphics processors were introduced, as well as low level TV color graphics for home use. Touchscreens were experimented with, light pens (a very good tool almost forgotten today) and voice command interfaces used. Music and animated films were computer generated. Top range Television sets got internal microprocessor control. This was truly an era of visionary ideas, many of which entice us even today.

2.7 The Event of the Cray-1

From 1968 till 1972 Seymour Cray was working on a successor to the CDC 7600—on the CDC 8600. It was a development of a scalar processor with many internal general purpose registers and the idea was to provide 4 of the "type 8000" processors in a single system with shared memory, allowing the same instruction to be executed by all four processors, effectively in a SIMD (Single Instruction Multiple Data) fashion. However, the overall complexity of the system regarding the available electronics and cooling technology was so great that no stable implementation could be developed, and in 1972 Mr. Cray stated to the Control Data Corporation management that the machine has to be fully redesigned from scratch. The management did not agree, Cray left the company and 2 years later, in 1974 the project was abandoned.

Instead of insisting further on the CDC 8600, the Control Data Corporation decided to make the STAR-100 [3]. Announced very early in 1970s, it first shipped in 1974, but its performance was much lower than hoped for and expected. The architecture was a fully pipelined Reduced Instruction Set (RISC) design, but the main and driving architectural decision was to make a vector processor (STAR comes from STring ARrays) which would have indeterminate vector lengths (i.e. could process vectors of any length up to 65,535 elements), and therefore was a memory to memory architecture. Unfortunately, the vector pipelining latencies allowed it to gain speed only on long vectors, whereas the performance on scalars was heavily strained by the vector-oriented approach. The memory to memory architecture necessitated a lot of memory accesses, as for example 5 consecutive operations on the same vector needed 10 memory accesses (5 reads and 5 writes) per element. Generally CDC STAR-100 was a disappointment, although it opened important insights into further avenues of computer architecture development.

After the experience with the unmanageable implementation of the Control Data Corporation CDC 8600 supercomputer, Seymour Cray decided to start his own company and design his fully own supercomputer, the Cray-1. However, he also learned a lot from the unsuccessful CDC STAR-100 design. It was important to make a machine which would be effective both in vector (even better than the STAR-100) and in the scalar (as the 8600) calculations. It is important to note that due to Amdahl's Law, the speed of the overall algorithm, whose one part is strictly serial and another is parallelizable (or, in this case, vectorizable), is directly dependable on the speed of the serial processing. Though this law is primarily meant for parallel computer architectures, it was shown by the CDC STAR-100 that just vector processing without speedy scalar and instruction series processing actually does slow down the overall throughput quite considerably.

As a result of these experiences Seymour Roger Cray made a fully new and a quite different design [3]. With 12.5 ns cycle time (memory and instruction), vector, scalar, address, base and table registers and independent functional units in three major segments (vector, scalar, address), full pipelining, instruction caching and vector chaining, the Cray-1 was far the fastest supercomputer of the 1970s. If programmed well, it could theoretically momentarily execute up to around 13 different instructions at the same time. However, sustained vector processing would normally allow between 2 and 4 concurrent instruction executions. Depending thus primarily on the program quality, the skill of the programmer and the algorithm, the Cray-1 of 1976 executed between 80 and several hundred million instructions per second (compare that 64-bit power with the meager 250-300 thousand 8-bit instructions per second of the MOS 6502.

The Cray computer series continued in development, and the successors of the Cray-1 design were for more than 25 years the fastest supercomputers in the world.

3 From History to Present

In the present it seems that we started forgetting the lessons learned and the design paths and avenues explored during the history of computers. In the 2014 Kailath lecture, prof. Donald Knuth stresses this point: "*For many years the history of computer science was presented in a way that was useful to computer scientists. But nowadays almost all technical content is excised;... We no longer are told what ideas were actually discovered, nor how they were discovered, nor why they are great ideas. We get only a scorecard.*" [4]. It seems that the perceived rapid development of computer technology, though in this last decade more perceived than real, stimulates in the minds of people a kind of "we are the best" feeling, resulting often in low regard towards prior computer systems and the perception that the modern computers are in all regards superior to the older ones, which is a highly suspicious claim.

One of the sad stories of Computer History is that there is in the world no existing Cray-1 series computer, only some cabinets and a few individual boards are preserved. Even worse, there is no known existing example of the original Cray Assembly Language (CAL) translator software, as well as no compilers, as it is true for (almost) all Cray-1/Cray-XMP series software, specifically development environment(s). The only known existing Cray-XMP software is a working Cray Operating System (COS), however without any useful programs (although there is an editor) or any translation software. COS was retrieved using the digital magnetic scanning method on a non-working hard disk from the Cray-XMP age and subsequent decoding of the oversampled files and finding out the data records format [5]. This amazing effort was done by Chris Fenton, the original author of the Verilog Cray-1 processor description, and Andras Tantos, the author of a fully-fledged Cray-XMP simulator written in C++ [6, 7]. The story of this heavy task of Digital Archeology is well worth reading about. The finally fully recovered COS necessitates a full single processor Cray-XMP system with the appropriate XMP IOPs (Input/Output Processors) and DMA (Direct Memory Access) (Fig. 2).

Lucky for History, a huge amount of documentation of the Cray processors is preserved by Bitsavers [8]. However, there are no technical plans, schematic diagrams, formulas. Therefore only a resurrection, reimplementation of black-box Cray processors from user and maintenance level documentation is possible. Such a re-implementation was initially done, as said, by Chris Fenton in Verilog [6] (which is actually a "more modern" version of what Mr. Cray did on paper in Boolean formulas—the Cray-1 system was not designed with circuit diagrams, but using a mathematical language). When the COS-recovery finally succeeded [9]

Fig. 2 Image of a historic Cray-1 supercomputer system. Image obtained from [12]

an open-source project on "Google Code" was started by Fenton and Tantos, with the aim to re-implement a full single processor Cray-XMP to be able to boot the resurrected Cray Operating System (COS) [7]. This point marked a significant split between the E2LP implementation and the Fenton/Tantos implementation. To implement the whole Cray-XMP system a significantly larger FPGA than the one on the E2LP board is necessary. The main aim of Fenton/Tantos implementation is a fully working system, regardless of the processor speed and efficiency, whereas the E2LP board implementation is primarily concerned with architectural design study and Cray architecture enhancements and modifications easily achievable with the Spartan 6 FPGA embedded on the E2LP board. The Fenton/Tantos implementation is geared towards resurrecting a fully working Cray-XMP system with original software, and not towards the development of new software development tools, whereas the educational intention of the E2LP implementation necessitates the possibility of programming the Cray processor, therefore at least a bootloader and an Assembly language translator were necessary to be developed.

Fortunately the same documentation preserved by Bitsavers.org allows the rebuilding of all necessary Cray-1 series basic software support.

4 The Cray-1 Architecture

The Cray-1 is an integrated address, scalar and vector processor. The whole processor can be regarded generally as having four parts: the instruction unit (the buffers, sequencer and decoder); the address processing unit; the scalar processing unit and the vector processing unit. It is designed in such a way that each active element is a self-standing and largely independent functional unit. This is true as for the mathematical and logical operation units as well as for the register files units and the memory interface unit. The Cray-1 design is very well thought out, very precise and very elegant. The computer is fully synchronous, and the clock frequency in 1976 was 80 MHz, giving a 12.5 ns instruction issue time.

A 64-bit word can represent either data or a part of an instruction stream. Basically the instructions themselves are largely 1 parcel wide (Cray name for 16-bit words, actually a parcel is the width of a simple instruction, the CDC 7600 word was 60-bits, so the parcel was 15 bits wide), but there are also 2 parcel wide instructions, for program jumps and absolute address memory fetches/stores. There are two types of addresses, which must not be intermixed: the program address, which is 16-bit (parcel) aligned, and the data address, 64-bit aligned. However, the memory can be fetched only by full words (64-bits). The instruction buffer logic fetches from memory in a burst 16 64-bit memory words into the least recently used of the 4 instruction buffers, therefore providing 64 single-parcel instructions. Each of the four instruction buffers has a base address register, which all are compared with the Program Counter for an instruction fetch. If none of the buffers contains the necessary instruction range, a buffer is filled again. The buffer fill starts with the Word containing the requested Parcel, so processing can

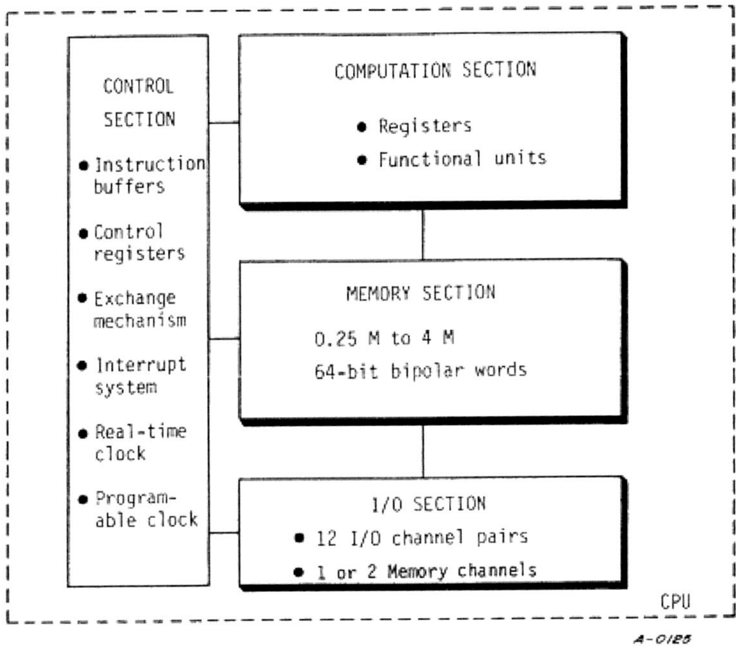

Fig. 3 Basic organization of the Cray-1 CPU. Image obtained from [13]

continue as soon as this Word is fetched from memory to the Instruction Buffer. Programs or program parts which can fit into 256 instructions will be executed exclusively from the instruction buffers, therefore not interfering with the main memory. This approach allows up to 4 program segments with nonsequential addresses and appropriate jump instructions to be kept in the processor (Fig. 3).

The Cray-1 is a three address machine, which means that each register oriented instruction (except for special purposes) has two source and one destination address (or a source and a destination address for monadic operations). The later Cray-XMP provided also two memory reads and one memory write per cycle, which would then be chained through the registers and computational units. To fit in the 16 bits of a one-parcel instruction, there are three register addresses of 3 bits each, allowing source and destination addressing of 8 registers of a certain type (address, scalar, vector). The remaining 7 bits are the appropriate instruction code. It is important to note that octal representation is used throughout the Cray documentation.

Due to pipelining and independent functional units, most of the instructions will actually need just one (decoding) cycle, to start the processing. Naturally using, for example, the same register before the result of the operation targeting that register is finished will stall the instruction sequence until the data is ready. To enable proper functioning and inter-synchronization of all the pipelines, all units have in all circumstances to have the same proper latency, and this "reservation time" is used to time-align the result with the appropriate register. The consequence of this is that a result of a short operation may arrive to its destination register before a result of a previously issued instruction which takes longer to execute. However, due to the pipelining of the destination register address and the reservation flags for each register and pipeline stage, the execution sequence stays the same as the program sequence, although further instructions can be processed in parallel from/to non-reserved registers. Another interesting consequence of having to know the execution latencies during the design stage of the hardwired processor is the lack of the DIVIDE instructions, both integer and floating-point. Instead a Reciprocal Approximation instruction is provided, as the algorithm has a predictable execution time in cycles, opposed to the division algorithm, which can take shorter or longer depending on data. Throughout the design such specific architectural decisions and optimizations come up.

The address calculation unit consists of 8 24-bit address (A) registers, 64 24-bit base (B) registers, an address add and an address multiply unit. Data can be transferred directly from registers A to B and vice versa, and any number of B registers in a sequence starting at any B register can be read from memory or written to it in a burst. Such a burst read or write will not interfere with any further instruction executions, as long as the source or destination of those instructions is not the main memory (or mostly even then, in the case of e.g. Cray-XMP), or the B register file itself. Any of the B registers can also be used with the Jump instruction, providing a program address. There is a possibility to transfer data between the A and S (scalar) registers. The address add and multiply units are independent and pipelined, so for example it is possible to load the B registers from memory, perform an address multiplication and perform an address addition in the same time (Fig. 4).

The Cray-1 has no subroutine stack; there is only a Return Jump instruction, which will save the return address (the address after the R instruction) into the base register B00. It is up to the programmer to take this into account if calling subroutines from subroutines!

The scalar calculation unit consists of 8 64-bit scalar (S) registers, 64 64-bit table (T) registers and 4 functional units: the Integer Add, Logical, Shift and Population/Leading Zero Count. There is no scalar integer multiply provided, address multiply may be used for smaller integers, and floating-point must be used for larger numbers. The shifts can be performed left or right by up to 63 bits at once, and 128-bit shifts are also provided. The population count and leading zero count are very useful operations normally not found on lower-end processors. The T registers behave analogous to the B registers, i.e. they provide a storage area for scalar data inside the processor. However, there is no special purpose T register like the B00 used for the return address.

The Cray-1 has no condition flags which would show the result of a most recent operation (like negative, overflow, zero etc.), as it is usual in many processors, primarily microprocessors. The conditional branch management using result condition flags is very complicated regarding pipelined architectures, as the condition flags can be set only at the end of the functional unit processing, which leads

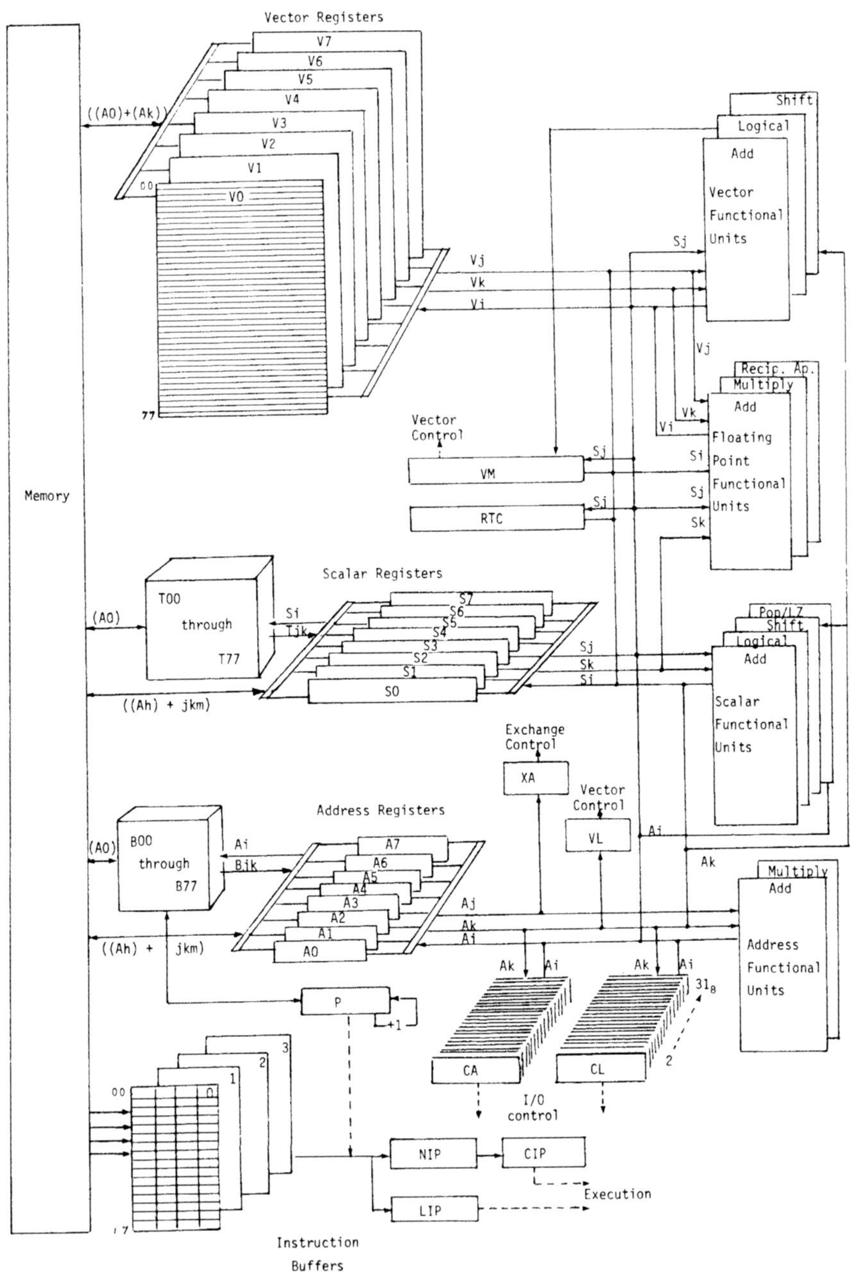

Fig. 4 Control and data paths in the Cray-1 CPU. Image obtained from [13]

to instruction issue stalls until the result is ready. The Cray solution is to use an address register (A0) and a scalar register (S0) which are always tested by A or S conditional instructions. This allows the programmer to insert a transfer to A0 or S0 in a proper place before the conditional Jump, or to prepare the conditions of a jump much in advance.

The vector calculation unit consists of 8 64-element 64-bit vector registers, a memory control functional unit and vector Integer Add, Logical and Shift units. Each of those four can function fully in parallel, that is a vector register can be interfaced to memory (read or written), and other vector registers can be processed by adding, logical operations or shifts. The later developed Cray-XMP, as mentioned, had the possibility to make two memory reads and one memory write to/from independent vector registers, in addition to the usage of other vector units. A Vector Length (VL) register (transfer with A file) is provided to define the processing length between zero and 64 elements. When issuing a vector instruction the vector length is remembered by the three addressed registers, so for subsequent instructions the length may be changed without interfering with the already started processing. A special vector shift is also provided shifting bits up to 63 places left or right taking the full vector length as a continuous string of bits. This means that at maximum a shift of 63 places can be done on a 4096-bit register (i.e. the 64 elements of 64-bits of a specified vector register pair). The shift count is defined by an A register. A 64-bit Vector Mask (VM) register is used for vector comparisons and vector merges. Memory vector accesses are defined by their base address (always in A0) and their stride (defined by any A), both defined in the vector memory instruction. The Vector Length is, as said, a separate register.

64-bit floating-point operations (in a Cray specific FP format) are provided by the FP functional units: Add, Multiply and Reciprocal Approximation. These functional units are shared between the scalar and the vector part of the processor. This sharing means that for example only integer operations and floating-point addition can be used on scalars while two vectors are being floating-point multiplied, and a vector's reciprocal approximation calculated. Otherwise the instruction sequencing will stall until the vector has finished processing (i.e. the FP unit will be ready its latency time before the last vector element is written to the result vector register—this allows full utilization). As scalars are single elements, a vector instruction using the same FP unit in the next instruction after a scalar FP operation will continue immediately. Proper organization of the code can mean a vast improvement of execution speed on the Cray, and special programs were developed to optimize the instruction sequences (e.g. SARA). However such software is also lost, destroyed or, if preserved, not publicly available.

To gain as much speed from using vector registers (as opposed to the memory to memory architecture of CDC STAR-100, discussed in the Historical overview) the vector instructions, i.e. the registers, can be "chained", thus the result of a calculation is simultaneously written to the register and forwarded to the functional unit stated in the next instruction. So for example, using the Cray-XMP (because of the advanced memory system), the following sequence would be executed for a data element each cycle (after initial latencies of all instructions): read V0 from memory, read V1 from memory, FP multiply V0 and V1 to V2, FP add V2 and S1 (scalar) to V3, make a FP reciprocal approximation of V3 to V4, write V4 to memory. After the added latencies of all these instructions, 6 operations are performed each cycle. However, this sequence will not influence the parallel execution of a vector addition of for example V5 and V6 into V5 and the shift of V5 by 17 places to V7, in addition to leaving much time to do scalar and address calculations in the same cycles.

5 CRAY Processor Implementation on the E2LP Platform

As already mentioned, the initial generic FPGA design of the Cray processor was recently developed by Chris Fenton, written in Verilog, and is in Public Domain. The implementation into the E2LP project involved the adaptation of the generic design to specific requirements to be used as a teaching tool, adaptation of the timings appropriate to the E2LP board, debugging of the original design, development of a Cray Assembly Language translator and development of a basic bootloader program for the E2LP Cray implementation.

During the architectural investigation and implementation on the E2LP board the internal timings were adapted to take advantage of the very high speed FPGA internal dual port memory, including the independent expansion of the original Fenton Cray-1 design with most of the Cray-XMP features. As there is no existing Cray software which could be used on a Cray-1, there was no need to strictly distinguish the Cray-1 from the Cray-XMP features, nor to implement certain instructions, like the channel instructions or the XMP shared registers. These exercises are left to the prospective students learning from this architectural implementation.

Due to the fact that the modern day DDR2 memory provided by the E2LP board has a reasonably complex interface, and is actually a long latency burst memory, necessitating memory-not-ready cycles in all chained registers and functional units (all others must be allowed to continue), the DDR2 memory is not used in the presented E2LP implementation, but this very interesting and deep problem is left for the students to experiment on their own. It must be noted that the 1976 Cray-1 did have memory-not-ready and memory-conflict wait cycles. The memory of the Cray-1 was banked in such a way that full throughput was achievable, except in stride vectors and memory bank conflict random accesses, or, later on the XMP, in memory banking conflicts between the doubled read and write memory access channels.

The E2LP implementation therefore has only 24 KiW (192 KiB) of memory, which is actually quite a large amount for experimental programming in assembly language. Several memory locations towards the end of the address range are reserved for E2LP I/O—a full UART is part of the E2LP-Cray system. The UART takes and produces full 8 byte, i.e. 64-bit, words, but features also a "character string length" register by which the transmitter/receiver can be turned off, or up to 8 characters expected/transmitted. Furthermore the 8-bit E2LP LEDs and

Table 1 E2LP board peripherals that are memory mapped on the Cray computer

Device/signal	Memory address range	Description
System RAM	0x000000–0x000FFF	Read or write
UART Tx Busy	0x100000	Read only
UART Rx Rdy	0x100001	Read only
UART Rx Data	0x100002	If read from this address
UART Rx Clr	0x100002	If written to this address
UART Tx Data	0x100003	Write only
UART CVL	0x100004	Character vector length: 0..8
LED output (LEDs 0..7)	0x100008	Read or write (8-bit)
Switches (0..7)	0x100009	Read only (8-bit)
Buttons (0..4)	0x10000A	Read only (5-bit)

Table 2 Resources on the Spartan-6 FPGA required to implement the Cray-1 design

Resource	Used	Available	Fraction (%)
Number of Slice Registers	8563	54576	15
Number of Slice LUTs	20169	27288	73
Number of fully used LUT-FF pairs	5158	23574	21
Number of bonded IOBs	28	358	7
Number of Block RAM/FIFO	105	116	90
Number of BUFG/BUFGCTRLs	1	16	6
Number of DSP48A1s	9	58	15
Number of PLL_ADVs	1	4	25

the push-buttons are memory mapped, to allow simple control. Table 1 gives the complete reference of all E2LP peripherals and their respective memory mapped locations.

The bootloader program is preloaded into the FPGA configuration bitstream, which allows it to execute immediately after the configuration, and/or system reset. It communicates with the external world through the serial line, and as such, the E2LP-Cray system behaves like any computer with a terminal connection. The bootloader presents the user with a simple interface, which enables loading a user program (in hexadecimal representation), resetting the program loading address, and initiating execution of the loaded program.

The synthesized design of the Cray-1 computer occupies a large amount of resources available on the Spartan-6 XC6SLX45 FPGA, as can be seen from Table 2.

Finally, a great introduction to the Cray-E2LP implementation, and a demonstration of its operation including the depiction of signal waveforms can be seen in an introductory video at https://www.youtube.com/watch?v=FltPGKWiEKI [10].

6 Conclusions

The Cray-1 implementation facilitates a perfect learning setup for students of all levels. It can guide a student from the very basic stages which involve the synthesis and transfer of the Cray-1 design onto the E2LP board up to the embedded software design in a real, comprehensive, and historically industrially very significant Cray Assembly Language [11]. As with any other CPU/MCU system, in order to gain the knowledge required to program the Cray-1 the student needs to get deeply familiar with its hardware architecture. With the wealth of documentation readily accessible, the students have an incredible opportunity to learn how one of the most powerful families of supercomputers operated for decades. Furthermore, some of the designs present in Cray-1 were significantly ahead of its time for widespread adoption, such as the vector processing units, which started to gain more popularity only a decade or so ago.

Additionally, many advanced exercises can be made with the core Cray processor implementation on the E2LP board. The expansion of the Cray-1 design into a Cray-XMP, Cray-2 or some other computer from that series enables deep insight in the correspondence of instruction sets, registers and interdependent timings. Connecting the board's DRAM to the core Cray processor enables the student to study the possibilities of banking, and the problems involved in coordination of the access times, asynchronicity and internal chaining and pipelining. Adapting the instruction set, implementing the IEEE floating point, rearranging the register sets and their access principles etc. are extremely important exercises to those who intend to work in processor design and construction, be it from the hardware, be it from the software side. Downscaling the Cray processor to e.g. 32-bits may show merits for special applications. As the data highways in the Cray processor are all 64-bit, and have to connect, inter alias, every vector register unit with every vector functional unit, and the memory functional unit with the floating point functional units, plus all the coordination buses, downgrading to 32 bits would produce a significantly smaller processor, giving much additional space on the E2LP FPGA. Finally, as the Cray processor takes up a large portion of the available FPGA space, students, when expanding the design, will be made aware of the timing and spacing limitations of a chosen FPGA, the possibilities and limitations of FPGA synthesis/implementation technologies, and will get involved in optimizations of different parts of the overall design, timing and other FPGA parameters.

An important aspect for the usefulness of the Cray processors for teaching purposes is that the documentation of the Cray designs preserved (by Bitsavers.org) is very thorough and extensive, therefore a full understanding and implementation is possible. Cray processors are completely hardwired (i.e. not microcoded) and fully synchronous, and therefore an excellent example of efficacy and the complexity/performance tradeoffs of computer design.

Acknowledgments The research leading to these results has received funding from the European Union's Seventh Framework Programme (FP7), under grant agreement no. 317882—*E2LP Embedded Computer Engineering Learning* Platform and Horizon 2020 Programme Integrating Distributed data Infrastructures for Global Exploitation—*INDIGO Data Cloud*.

References

1. Hwang, K., Jotwani, N.: Advanced Computer Architecture: Parallelism, Scalability, Programmability. Tata McGraw Hill Education Private Limited, New Delhi (2011)
2. Ceruzzi, P.E.: A History of Modern Computing. MIT Press, Cambridge (2003)
3. Oklobdzija, V.G.: The Computer Engineering Handbook. CRC Press, Boca Raton (2001)
4. Knuth, D.: 2014 Kailath Lecture: Stanford Professor Donald Knuth. http://www.youtube.com/watch?v=gAXdDEQveKw (2014). Accessed July 2015
5. Fenton, C.: Cray-1 Digital Archeology. http://www.chrisfenton.com/cray-1-digital-archeology/. Accessed July 2015
6. Fenton, C.: Homebrew Cray-1A. http://www.chrisfenton.com/homebrew-cray-1a/. Accessed July 2015
7. Fenton, C., Tantos, A.: cray-1x—an FPGA-based implementation of the Cray-1 Supercomputer. http://code.google.com/p/cray-1x/. Accessed July 2015
8. "Cray". http://bitsavers.org/pdf/cray. Accessed July 2015
9. Fenton, C.: COS Recovery. http://www.chrisfenton.com/cos-recovery/. Accessed July 2015
10. Šojat, Z.: Reincarnation CRAY-1 on E2LP Platform. https://www.youtube.com/watch?v=FltPGKWiEKI. Accessed July 2015
11. Rogina, B.M. et al.: Exercises for Embedded Engineering Learning Platform, chapter in Springer special edition book: Embedded Engineering Education, in this book
12. "Cray-1". http://www.cray.com/sites/default/files/resources/CRAY-1.jpg, Accessed July 2015
13. "CRAY-1 Computer System Hardware Reference Manual 2240004". http://www.textfiles.com/bitsavers/pdf/cray/2240004C-1977-Cray1.pdf, Accessed December 2015

Methods for User Involvement in the Design of Augmented Reality Systems for Engineering Education

Margarita Anastassova, Sabrina Panëels and Florent Souvestre

Abstract The paper presents a number of user involvement methods which can be used in the design of Augmented Reality (AR) systems for engineering education. One of the characteristics of these technologies is that future users do not always have a thorough knowledge of AR and its applications in engineering education. Furthermore, the technology is in search of applications, and there are few existing HCI guidelines for AR interfaces. In this sense, the design and usability evaluation of these systems are real challenges. We present methods which are suitable in this context (e.g. scenarios, field studies, activity analysis, and formative evaluations of prototypes). We also discuss their advantages and limitations when designing AR systems for engineering education from a user-centred perspective.

Keywords Augmented reality · User involvement · Methods · Evaluation · Requirements elicitation · Engineering education

1 Introduction

The term "Augmented Reality" (AR) was introduced in the early 1990s [1] to designate a specific form of Human-Computer Interaction (HCI), in which views of the real world are enhanced by computer-generated content [2]. The real and virtual elements in an AR system are semantically and spatially related. Compared

M. Anastassova (✉) · S. Panëels · F. Souvestre
CEA, LIST, Sensorial and Ambient Interfaces Laboratory,
91191 Gif-sur-Yvette Cedex, France
e-mail: margarita.anastassova@cea.fr

S. Panëels
e-mail: sabrina.paneels@cea.fr

F. Souvestre
e-mail: florent.souvestre@cea.fr

R. Szewczyk et al. (eds.), *Embedded Engineering Education*,
Advances in Intelligent Systems and Computing 421,
DOI 10.1007/978-3-319-27540-6_5

to Virtual Reality (VR), AR does not aim at representing the real world by a realistic virtual analogy. It aims at promoting “intuitive” and natural multimodal interaction [3]. In addition to 2D and 3D computer-generated visual content, spatial audio, tactile and even olfactory stimulations can be incorporated to enhance the user’s perception of the real world.

As underlined by [4], AR offers new possibilities in education. These authors, together with others, cite the following major advantages of using AR in education:

- a possibility of presenting information “just-in-time” and “just-in-place”, which will reduce information search, error-likelihood and will enhance memorization and recall (e.g. [5–7]);
- a possibility of visualizing complex relationships and abstract concepts ([8, 9]);
- a possibility of experiencing phenomena which are unlikely to be experienced in the real world ([10, 7]);
- a possibility of “learning-by-doing” (i.e. of constructing knowledge actively and autonomously, [11, 12]);
- a possibility of improving learners’ motivation because of the enthusiasm when interacting with new technologies ([13]).

All this benefits are transposable to engineering education, in general, and to embedded electronics courses, in particular.

There are only few AR reality prototypes for engineering education. Some of them will be presented below.

2 AR in Engineering Education

Kaufmann and Schmalstieg [14] developed an AR system for mathematics and geometry education (Fig. 1).

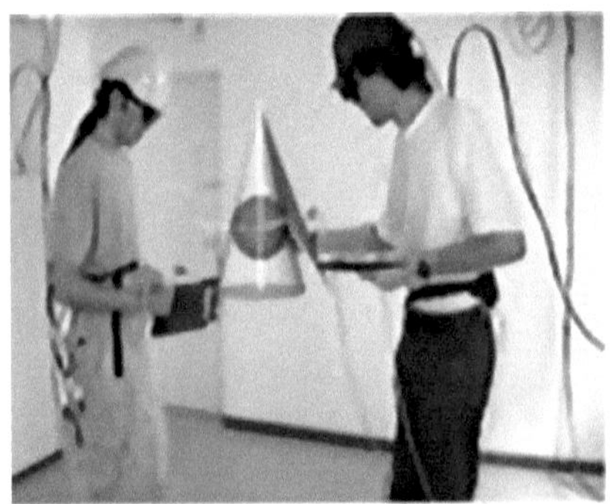

Fig. 1 AR system for geometry education (adapted from [14])

The system is a 3D geometric construction tool for the improvement of spatial abilities and for the maximization of transfer in real settings. This system has not been formally evaluated in a real course. However, an informal evaluation showed that students were motivated to use it and did not need a long familiarization before using it in practice. Several problems such as eye-hand coordination without haptic feedback and fatigue were also pointed out. As for the possible applications of the system, students mentioned interactive conic sections, vector analysis, intersection problems, and elementary geometry.

Another example is the use of tangible interfaces (i.e. physical objects coupled to digital information) and AR models in engineering graphics courses to help students better understand the relationship between 3D objects and their projections [15]. This system was tested with 35 engineering-major students. The study showed that the tangible interfaces significantly enhanced students' learning performance and their abilities to transfer 3D objects onto 2D projections. There was also high engagement with the AR models during the learning process.

AR was also used for teaching embedded electronics courses. When learning electronics, especially embedded systems, students have to face the challenge of understanding the mechanisms of several devices without actually seeing those interactions and functions. Even in laboratory practices with electronic boards, they can only manipulate them through the available inputs and outputs, whilst the operations happening inside the components remain invisible. Consequently, students do not always get to fully understand the studied concepts.

AR aims to overcome those obstacles in the learning process, especially in the early stages of Computer Engineering studies. Thus, Müller et al. [16] and Andujar et al. [17] proposed an AR system for the improvement of students' interactions with remote laboratories.

In [17], the use case is the design of a digital control system based on an FPGA development board. In this case, AR is used in order to give the user the sensation that certain lab functions can be handled just as they would be in the real laboratory itself. The authors designed the system with the aim of limiting students' possible discouragement due to the lack of physical contact. The system was evaluated with 36 students and 10 teachers. The results, both for students and teachers, showed improved learnability of the theoretical concepts taught in the different courses, high engagement and higher motivation to learn than with traditional methods.

3 AR: An Emerging Technology Which Can Benefit from Stronger User Involvement

AR is still an emerging technology and its potential utility for engineering education, though described in the literature, is not always clearly envisioned by its future users (students and teachers). This can be partially explained by the fact that this technology is not well-known by these users. The lack of knowledge and experience with the technology makes difficult the expression, collection, analysis and formalization of user needs and potential applications of AR. In the same time, this analysis is crucial for the design of products satisfying the future users' expectations [18].

The following benefits of strong user involvement in the design of emerging technologies such as AR have been stated in the literature:

- a clear definition of product and project objectives [19];
- a withdrawal of a number of costly features which can be unwanted or of little use for future users;
- better acceptability of the product [20];
- better understanding of the system resulting in a more effective use;
- a positive effect on the short and medium-term use of the system [21, 22];
- general user satisfaction and a higher level of perceived usefulness of the system [21, 23];
- better understanding and coordination between all actors involved in the design process;
- better users' knowledge of their own activity and of the ecosystem in which they evolve [24];
- a smaller number of iterations of the design cycle compared to projects with little or no user involvement [25].

One of the necessary conditions for obtaining these positive results is to involve users as early as possible in the design process [20]. However, an early user involvement approach is rarely adopted during the design of emerging technologies, including AR, and this for a number of reasons to be detailed below.

4 Difficulties for Early User Involvement in the Design of Emerging Technologies

An initial and essential difficulty lies in the originality of the emerging technologies, which explains the fact that they seldom meet "conscious" [26] or explicitly formulated user requirements. In many cases, these technologies are technical innovations, which meet "latent" [27] user requirements or simply create new ones. In other words, emerging technologies are often designed because designers know how to develop them and assume they will meet a user requirement [27]. Robertson [26] calls these types of user requirements "undreamed-of requirements" and states that it is difficult for future users to clearly express them.

Therefore, the mere identification of future users becomes problematic [28]. Even if emerging technologies can target a given group of potential users, the user characteristics and tasks are ill-defined at the beginning of the design process.

Also, rather often and for various reasons, designers of emerging technologies do not ask for early user involvement [29, 30]. Therefore, a majority of Human-Computer Interaction (HCI) endeavours are limited to more or less formal (possibly iterative) user evaluations of technological prototypes. Detailed user requirements analysis is not always performed in the very beginning of the design process.

User involvement difficulties may also arise from a number of "restrictive" [31] principles guiding designers' decisions (e.g. simplicity of the final solution, rapidity of development, realism [30, 32]. In a project, these principles may orient design efforts to a product which necessarily works, as soon as possible, but often out of its future context of use [23, 33].

We should also admit that the available HCI methods for user involvement do not always give the expected results when working on emerging technologies, as they usually describe user activities at a given point in space and time rather than anticipate them [27].

A number of commonly used methods for user involvement as well as their application to the design of AR systems will be presented below.

5 Methods for User Involvement in the Design of AR Systems

The methods and techniques for user involvement in the design of AR systems for educational or other applications can be organised in the following four broad categories:

- methods relying on state-of-the-art knowledge relevant for the future situation of use of a given emerging technology;
- scenarios and similar techniques portraying the emerging technology to its future users;
- formal and informal user evaluations of mock-ups and prototypes;
- user requirements extrapolation from actual users' activities [28, 34].

These four broad categories of methods and techniques for user involvement will be briefly presented below.

Methods relying on state-of-the-art knowledge relevant for the future situation of use of a given emerging technology: State-of-the-art knowledge available in research papers, guidelines and norms is based on user studies with existing technologies similar to the emerging technology to be designed. Using such ready-to-apply knowledge during the design process can be valuable for a number of reasons, namely:

- starting from application of existing technologies, designers can target possible applications of the emerging technology they are working on;
- provide a "database" of potentially useful and reusable features and functionalities [35];
- provide feedback on advantages and drawbacks of similar existing technologies [28, 36].

As far as AR for engineering education is concerned, there is limited specific ready-to-apply knowledge (e.g. detailed collection of design guidelines). Most of the existing guidelines are adapted from guidelines used for more traditional interactive systems and/or for Virtual Reality systems (e.g. [37]). In addition, a number of these guidelines are not empirically validated and rather generic [34].

These limitations of existing guidelines for AR systems for engineering education can be explained by the fact that the user interaction technologies in the field of AR are numerous and varied (e.g. head-worn displays, smartphones, traditional screens, sensors). For this reason and because of the limited user experience with these interaction technologies, user performance can be very disparate. In this sense, the creation of design guidelines, which are general, consistent and consolidated enough to be directly used by AR system designers is a difficult task.

Thus, designing AR systems for engineering education starting from state-of-the-art knowledge only may be challenging. Other methods for direct user involvement (e.g. scenarios of future system use [38] can enrich the design process of emerging technologies.

Scenarios and similar techniques portraying the emerging technology to its future users: Scenarios and similar techniques such as use cases allow designers to explicitly envision and document typical and significant user activities early and continuingly in the development process [38]. In this sense, they give a concrete representation of the actual use of a future product or system [39]. Scenarios can be presented as stories, storyboards, videos.

In the field of AR, short stories [40] and videos (the use case scenario for the The Magic Book, http://www.mic.atr.co.jp/~poup/research/ar/index.html) have been used. Videos are particularly adapted for AR, which is often based on visual or multimodal informational inputs.

Because of their concreteness, scenarios and use cases, in different forms, can facilitate a number of discussions on the functionalities of the future system [41] and thus help elicit latent or undreamed-of requirements.

However, when working on emerging technologies, these design tools have at least three limitations. First, they may be incomplete, since the future uses of the emerging technology are, by definition, ill-defined. Second, they can express predominantly the viewpoint of the person who created them [41]. Finally, the concreteness of scenarios and use cases can inhibit designers' imagination and creativity [42]. To overcome this limit, scenarios and use cases can be presented to users and designers in focus groups. This form of social interaction could be preferable because it allows open exchanges between different project stakeholders and helps generating innovative solutions [26].

For the same reason, focus groups can be used for more or less formal user studies with mock-ups and prototypes.

Formal and informal user evaluations of mock-ups and prototypes: User evaluations of mock-ups and prototypes can be either formal or informal. Formal evaluation is based on the rigorous application of experimental principles, whereas informal evaluation is often conducted without systematically controlling experimental factors. A paradigmatic example of informal user evaluation of

AR prototypes is their presentation to visitors and participants in congresses and conventions.

Originally, this method was used to assess user performance with a given technology. However, when designing emerging technologies such as AR, this method can be a valuable tool for (re)defining the potential applications of a system. Thus, the precise definition of user requirements for an AR system to assist architects during the exploratory phases of architectural design, was performed using several formal and informal evaluations of prototypes [43].

As for scenarios and use cases, the main advantage of mock-ups and prototypes as tools for user involvement is their concreteness. However, mock-ups and prototypes are physical representations of the future technology which can be easily manipulated.

In this sense, they can help the elicitation of undreamed of requirements and, in the same time, facilitate the identification of usability problems [35]. Also, prototypes are essential when designing sensory (e.g. haptic or olfactory) interaction.

Unfortunately, emerging technologies, because of their prototypical nature, may be technically or functionally immature and/or unstable. Furthermore, user experience with innovative technologies may be influenced by users' habits of interaction with more mature technologies. This influence is rarely in favour of immature prototypes. As a result, users may feel frustrated when interacting with prototypes of emerging technology and eventually reject the technology to be designed. For this reason, it is important to combine user studies with mock-ups and prototypes with other methods for user involvement (e.g. activity and task analysis).

User requirements extrapolation from actual users' activities: In this category includes methods such as interviews, direct observation of activities, task analysis and task modelling. Interviews can be a very valuable and relatively easy-to-use and cost-effective method for user involvement early in the design process. The method can give interesting results if the future users sample is large and varied enough, and if all the facets of the users' activities are explored.

Interviews and questionnaires are often used to inform the design of AR systems. Thus, in order to define the user interface and the functionalities of an AR prototype for military applications, navy officers were interviewed [44]. In the same vein, two refinery operators were interviewed early in the design of an AR system for the training of petrochemical engineers [45]. However, in these studies the user samples were quite limited, which influences negatively the validity of their results. In addition, as pointed by Van Schaik [46], in most cases, it is not enough to ask the future users of an emerging technology what they want since:

- users are focused on their day-to-day activity rather than on technology design [23].
- users do not necessarily specify the features of a future system in a terms which are directly usable by designers. They often point problems rather than technological solutions [46].
- users express numerous and heterogeneous requirements.

Interviews can be complemented by observations of users' activities and subsequent task analyses. These methods are particularly useful for analysing tasks which require implicit operational knowledge, which is often difficult to externalise. Consequently, activity observations can be a valuable tool to explicit latent or undreamed-of requirements [47].

However, as pointed by [28], actual activity analysis may limit the design of an emerging technology since it does not incorporate the changes which this technology will introduce into user's everyday tasks. Furthermore, observational studies are mostly descriptive since focused on current users' practices. In this sense, they are may be perceived as insufficiently predictive or too vague by designers [27]. Finally, the amount of the collected data and its analysis can be overwhelming. This finding is not new. Leplat [48] states that task analysis may be long and difficult since it helps precisely define the scientific problem in technology design. Still, these authors emphasize that pretending to solve a problem in this field without prior task analysis would mean prescribing medication to a patient without previously examining her/him.

6 Discussion and Conclusion

User involvement early in the design of emerging technologies such as AR is a challenge for HCI methodology and knowledge, since innovation is upcoming and in search of potential applications. Consequently, it is barely known by its future users. Thus, users are not likely to express their needs for innovation because they can hardly imagine and describe what might be possible to do with an eventual future technology. In addition, the more radical an innovation the harder it is to understand how it should look, function, and be used. In general, people are most prone to communicate needs which they are particularly aware of. Therefore, most of the HCI methods traditionally used for user needs analysis help the elicitation of such conscious user needs, thus undoubtedly informing design and key industrial stakeholders. However, during the early design of emerging technologies, users are required to express their undreamed of requirements [26], and unless people are encouraged explicitly to think about such requirements, they are unlikely to appear until later in the development of a technology, when its potential applications become clear and evident [27].

In consequence, all types of representations of the future technology (e.g. prototypes and scenarios) could be very useful in order to give users an idea about technological constraints and possibilities and, thus, to elicit their undreamed of requirements. Another useful technique could be literature analysis on the actual or envisioned applications of a given emerging technology. Such literature reviews would favour the construction of a needs database as well as the reuse of user requirements.

As for the user-evaluation of prototypes of emerging technologies, it is a challenge for HCI knowledge, because the existing prototypes propose a limited number of functions to support user activities (e.g. visualization, selection, positioning, and rotation of virtual entities in the case of AR prototypes). In addition, existing AR prototypes present several technological limitations such as registration errors. The latter limitations influence both the usability and the acceptability of any AR educational application. Nevertheless, this method might be a driving force for innovation because it encourages users' exploration. In this sense, in order to overcome the difficulties related to the low degree of fidelity of the prototypes and to better inform designers, focus groups or other relatively informal social settings may be used for prototype evaluation.

Another possible option is the comparison with task-based criteria instead of comparisons with a traditional, well-known and well-accepted technology.

In conclusion, we would like to emphasize the importance of more methodologically-oriented studies concerning both user-needs elicitation and prototype evaluation of AR for engineering education, since the results currently available are neither satisfactory, nor sufficient in order to establish definitive conclusions.

References

1. Caudell, T., Mizell, D.: Augmented reality: an application of heads-up display technology to manual manufacturing processes. In: 1992 Proceedings Hawaii International Conference on System Sciences, vol 2, pp. 659–669
2. Yuen, S.-Y., Yaoyuneyong, G., Johnson, E.: Augmented reality: an overview and five directions for AR in education. J. Educ. Technol. Dev. Exch. **4**, 119–140 (2011)
3. Billinghurst, M., Kato, H., Poupyrev, I.: The magic book: moving seamlessly between reality and virtuality. Comput. Graph. Appl. **21**, 2–4 (2001)
4. Wu, H.-K., Lee, S.W.-Y., Chang, H.-Y., Liang, J.-C.: Current status, opportunities and challenges of augmented reality in education. Comput. Educ. **62**, 41–49 (2013)
5. Anastassova, M., Burkhardt, J.M., Mégard, C., Ehanno, P.: L'ergonomie de la réalité augmentée: une revue. Le Travail Humain **70**(97–126), 2007 (2007)
6. Cooperstock, J.R.: Classroom of the future: enhancing education through augmented reality. In Smith, M.J., Salvendy, G., Harris, D., Koubek, R.J. (eds.) Usability evaluation and interface design: cognitive engineering, intelligent agents and virtual reality (pp. 688–692). Lawrence Erbaum Associates, Mahwah (2001)
7. Neumann, U., Majoros, A.: Cognitive, performance, and systems issues for augmented reality applications in manufacturing and maintenance. In: 1998 VRAIS Proceedings of the IEEE Virtual Reality Annual International Symposium, pp. 4–1 (1998)
8. Arvantis, T.N., Petrou, A., Knight, J.F., Savas, S., Sotiriou, S., Gargalakos, M., et al.: Human factors and qualitative pedagogical evaluation of a mobile augmented reality system for science education used by learners with physical disabilities. Pers. Ubiquit. Comput. **13**, 243–250 (2007)
9. Shelton, S., Hedley, N.: Using augmented reality for teaching earth-sun relationships to undergraduate geography students. In: Proceedings of 2002 1st IEEE International Augmented Reality Toolkit Workshop, Darmstadt, Germany, Sept 2002
10. Klopfler, E., Squire, K.: Environmental detectives: the development of an augmented reality platform for environmental simulations. Educ. Tech. Res. Dev. **56**, 203–228 (2008)
11. Doswell, J., Blake, B., Green, J., Mallory, O., Griffin, C.: Augmented Reality learning games: 3D virtual instructors in augmented reality environments. Paper presented at the 2006 Symposium on Interactive Games, 2006

12. Fjeld, M., Voegtli, B.: Augmented chemistry: an interactive educational workbench. In: Proceedings of the 2002 IEEE/ACM International Symposium on Mixed and Augmented Reality (ISMAR 2002), pp. 259–260
13. Zhong, X., Liu, P., Georganas, N., Boulanger, P.: Designing a vision-based collaborative augmented reality application for industrial training. IT Inf. Technol. **45**, 7–18 (2003)
14. Kaufmann, H., Schmalstieg, D.: Mathematics and geometry education with collaborative augmented reality. Comput. Graph. **27**, 339–345 (2003)
15. Chen, Y.-C., Chi, H.-L., Hung, W.-H., Kang, S.-C.: Use of tangible and augmented reality models in engineering graphics courses. J. Prof. Issues Eng. Educ. Pract. **137**, 267–276 (2011)
16. Müller, D., Bruns, F.W., Erbe, H.-H., Robben, B., Yoo, Y.-H.: Mixed reality learning spaces for collaborative experimentation: a challenge for engineering education and training. IJOE Int. J. Online Eng. **3**(4) (2007). http://www.informatik.uni-bremen.de/~mueller/en/publ_assets/2007-ijoe-mueller.pdf
17. Andújar, J.M., Mejías, A., Márquez, M.A.: Augmented reality for the improvement of remote laboratories: An augmented remote laboratory. IEEE Trans. Educ. **54**, 492–500 (2011)
18. Davis, A.M., Zowghi, D.: Good requirements practices are neither necessary nor sufficient. Requirements Eng. **11**, 1–3 (2006)
19. Nielsen, J.: Usability Engineering. Academic Press, London (1993)
20. Damodaran, L.: Human factors in the digital world enhancing life style—the challenge for emerging technologies. Int. J. Hum. Comput. Stud. **55**, 377–403 (2001)
21. Barki, H., Hartwick, J.: User participation and user involvement in information system development. In: Proceedings of the 24th Annual Hawaii International Conference on System Sciences, vol. IV, pp 487–492 (1991)
22. Baroudi, J.J., Olson, M.H., Ives, B.: An empirical study of the impact of user involvement on system usage and information satisfaction. In: Communications of the ACM (Ass. for Computing and Machinery), vol. 29, pp. 232–238 (1986)
23. Foster Jr, S.T., Franz, C.R.: User involvement during information systems development : a comparison of analyst and user perceptions of system acceptance. J. Eng. Tech. Manage. **16**, 329–348 (1999)
24. Wilson, A., Bekker, M., Johnson, H., Johnson, P.: Costs and benefits of user involvement in design: Practitioners' views. In: Proceedings of HCI'96, People and Computers XI (London: Springer Verlag), pp. 221–240 (1996)
25. Chatzoglou, P.D., Macaulay, L.A.: Requirements capture and analysis: a survey of current practice. Requirements Eng. **1**, 75–87 (1996)
26. Robertson, S.: Requirements trawling: techniques for discovering requirements. Int. J. Hum. Comput. Stud. **55**, 405–421 (2001)
27. Sperandio, J.C.: Critères ergonomiques de l'assistance technologiques aux opérateurs. In: Communication présentée à JIM'2001: Interaction Homme—Machine & Assistance, Metz, France, July 2001
28. Brangier, E., Bastien, J.M.C.: L'analyse de l'activité est-elle suffisante et/ou pertinente pour innover dans le domaine des nouvelles technologies? In: Valléry, G., Amalberti, R. (eds.) L'analyse du travail en perspectives. Influences et évolutions. Toulouse, Octarès 2006
29. Flores, F., Graves, M.: Computer systems and the design of organizational interaction. ACM Trans. Off. Inf. Syst. **6**, 504–513 (1988)
30. Følstad, A., Rahlff, O.W.: Challenges in conducting user-centred evaluations of mobile services. Paper presented at HCII 2005 Human-Computer Interaction International Conference, Las Vegas, NV, USA, July 2005
31. Grundin, J.: Systematic sources of suboptimal interface design in large product development organizations. Hum. Comput. Interact. **6**, 147–196 (2001)
32. Kjeldskov, J., Graham, C.: A review of mobile HCI research methods. In: 5th International Symposium, Mobile HCI. Lecture Notes in Computer Science: Human-Computer Interaction with Mobile Devices, pp. 317–335. Springer, Berlin, Heidelberg (2003)

33. Stary, C.: Shifting knowledge form analysis to design: requirements for contextual user interface development. Behav. Inf. Technol. **2**, 425–440 (2002)
34. Burkhartd, J.M.: Ergonomie, facteurs humains et réalité virtuelle. In: Fuchs, P., Moreau, G., Berthoz, A., Vercher, J.-L. (eds.) Le traité de la réalité virtuelle (pp. 117–150). Presses de l'École des Mines, Paris (2006)
35. Robertson, S., Roberston, J.: Mastering the Requirements Process. Addison-Wesley & ACM Press, New York 2006
36. Maguire, M., Bevan, N.: User requirements analysis: a review of supporting methods. Paper presented at IFIP 17th World Computer Congress, Montreal, Canada, August, 2002
37. Bach, C.: Élaboration et validation de critères ergonomiques pour les interactions Homme—Environnements Virtuels. Ph.D. thesis, Université de Metz (2004)
38. Carroll, J.M.: Scenario-Based Design: Envisioning Work and Technology in System Development. Wiley, New York (1995)
39. Carroll, J.M.: Making Use: Scenarios and Scenario-Based Design of Human-Computer Interactions. MIT Press, Cambridge (2000)
40. Balaguer, A., Lors, J., Junyent, E., Ferr, G.: Scenario based design of augmented reality systems applied to cultural heritage. Paper presented at the Panhellenic Conference on Human Computer Interaction, Patras, Greece, December 2001
41. Sutcliffe, A.: User-Centred Requirements Engineering. Springer, Berlin (2002)
42. Lindgaard, G., Dillon, R., Trbovich, P., White, R., Fernandes, G., Lundahl, S., Pinnamaneni, A.: User needs analysis and requirements engineering: theory and practice. Interact. Comput. **18**, 47–70 (2006)
43. Aliakseyeu, D., Martens, J.B., Rauterberg, M.: A computer support tool for the early stages of architectural design. Interact. Comput. **18**, 528–555 (2006)
44. Gabbard, J.L., Hix, D. Researching usability design and evaluation guidelines for augmented reality systems, Students Project. (www.sv.vt.edu/classes/ESM4714/Student_Proj/class00/gabbard/). (2001)
45. Träskbäck, M., Haller, M.: Mixed reality training application for an oil refinery : user requirements. Virtual Reality Continuum and its applications in Industry. Paper presented at VRCAI 04, Singapore, June 2004
46. van Schaik, P.: Involving users in the specification of functionality using scenarios and model-based evaluation. Behav. Inf. Technol. **18**, 455–466 (1999)
47. Kujala, S.: User involvement: a review of the benefits ans challenges. Behav. Inf. Technol. **22**, 1–16 (2003)
48. Leplat, J.:15 ans d'analyse de l'activité: quelles évolutions? In; Vallléry, G., Amalberti, R. (eds.) L'analyse du travail en perspectives. Influences et évolutions. Toulouse, Octarès 2006

Augmented Reality Interface for E2LP: Assistance in Electronic Laboratories Through Augmented Reality

Enara Artetxe González, Florent Souvestre and Jorge R. López Benito

Abstract This study presents an Augmented Reality Interface for engineering education. The interface, designed to use Augmented Reality to facilitate learning, is composed of both specific software and hardware elements and provides useful information and assistance in Electronic Laboratories. The document first presents the overall system and its objectives under the E2LP project. Then its components, their functioning and adaptation to educational purposes are discussed in detail. The study concludes with the approach of the scalability of the system and its future use in classrooms.

Keywords Augmented reality · Augmented reality interface · Electronic laboratories · Engineering · Education

1 Introduction

Augmented Reality (AR) consists of the combination of the real world with virtual elements through a camera in real time. This emerging technology has already been applied in industrial fields such as production and maintenance with several benefits, e.g. time reduction to locate and perform a task, improvement of the learning process and increment of overall efficiency [1, 2].

E. Artetxe González (✉) · J.R. López Benito
CreativiTIC Innova S.L, La Rioja, Spain
e-mail: eartetxe@creativitic.es

J.R. López Benito
e-mail: jrlopez@creativitic.es

F. Souvestre
CEA, LIST, Sensory and Ambient Interfaces Laboratory,
Gif-Sur-Yvette Cedex, Paris, France
e-mail: florent.souvestre@cea.fr

R. Szewczyk et al. (eds.), *Embedded Engineering Education*,
Advances in Intelligent Systems and Computing 421,
DOI 10.1007/978-3-319-27540-6_6

In education, similar approaches have been taken in order to improve the comprehension of abstract concepts such as electronic fields, and enhance the learning process making education more interactive and appealing for students [3, 4].

One of the main innovations of the ICT FP7 project E2LP [5] consists of an Augmented Reality Interface (ARI) that detects different electronic boards and superposes relevant information over their components, serving also as a guide through laboratory exercises [6].

This document presents the main features of the ARI, the necessities it covers, its development and integration in the E2LP platform.

2 Objectives

The main objective of the ARI is to provide students and teachers from Electronic Laboratories a support tool that will provide useful information about the boards and exercises through a user-friendly interface and using Augmented Reality technology. Sub-objectives of this point are the *scalability* and *adaptability* of the tool, providing easy access to the further introduction of contents and features in the system and allowing its future expansion beyond the scope of the project.

From the educational point of view, being encompassed in the general scope of the E2LP project, the ARI has been designed according to the task taxonomy established within the project [7], providing a wider range of AR utilities for Basic Exercises and serving as a support tool for the rest of the exercises, i.e. Problems and Projects (see Table 1).

3 ARI: The System

The ARI system (Fig. 1) consists of:

- An articulated arm with a touchscreen and a webcam attached.
- A multi-feedback pointer.
- A mini-PC integrating the logic of the system.

Table 1 Classification of augmented reality actions according to task taxonomy

No	Augmented reality actions in task taxonomies			
	Action name[a]	Ex	Pb	Pr
1	Displaying of components' datasheets when touching them	X	X	X
2	Concept introduction: explanation of theoretical concepts, theories etc	X	X	X
3	Highlighting of the hardware components to be used in an exercise	X	X	X
4	Displaying of steps or instructions that have to be followed to successfully complete the task	X	X	
5	Explanation of the solution or some of the possible solutions directly over the components of the board	X	X	

[a] *Ex* Exercises, *Pb* Problems (open-ended tasks), *Pr* Projects (student-defined open tasks)

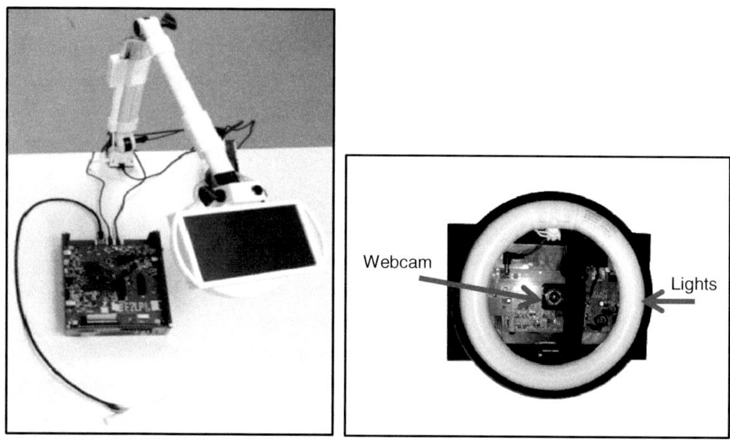

Fig. 1 Augmented reality interface overview (*left*) and magnifying glass bottom view (*right*)

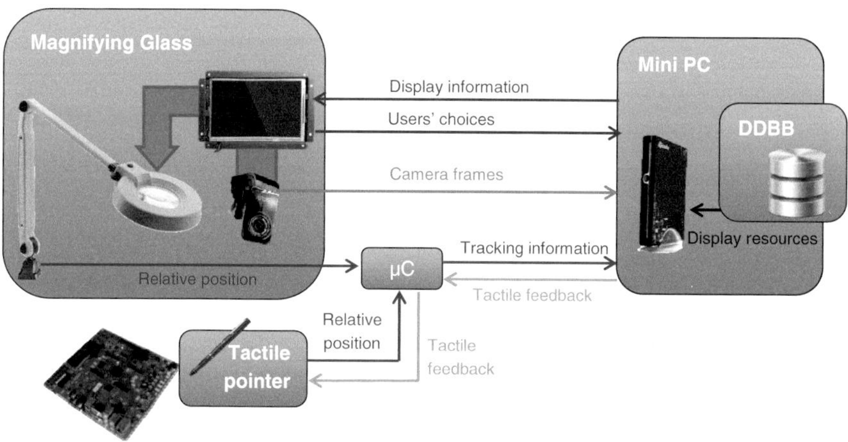

Fig. 2 Functional diagram of ARI

The schema in Fig. 2 shows the main connections and data exchange between the elements of the ARI.

The mini-PC receives the camera frames and the position from the magnifying glass' sensors and tactile pointer. It processes the information against its database to display in the touchscreen the augmented data.

Users interact with the touchscreen to select components, exercises and scenarios.

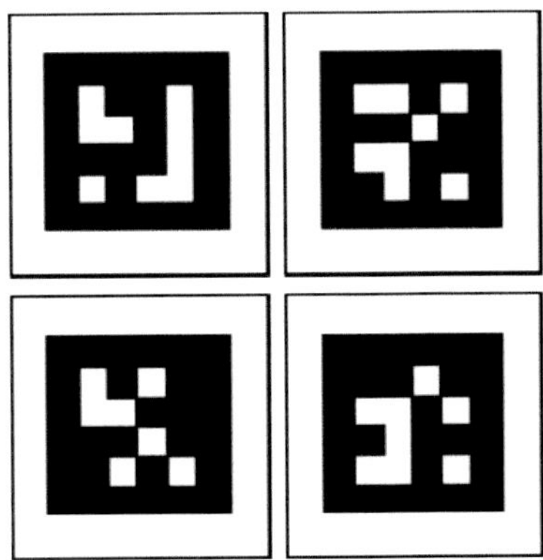

Fig. 3 The chosen markers for the E2LP board

4 ARI Components: Software

The software of the ARI is divided in two sections: the tracking software that allows augmented content to be displayed; and the users' interface to interact with the system and introduce new exercises.

4.1 AR Tracking Software

The tracking software of the ARI has been developed using **OpenCV** [8], an open source computer vision library which has all the necessary capacities for the picture processing and tracking [9]. This library is fully compatible with **OpenGL** [10], the open graphics library that allows the creation of the augmented reality layer, displaying the required information over the video stream layer, and that will be used due to its standardization and multiplatform nature.

For the software tracking, the task has been divided in two sections: a main process based on the image-tracking (**markerless**) of the board and a secondary **marker-based** tracking, where small patterns (Fig. 3) have been placed in the main E2LP board.

4.1.1 Markerless Tracking

Image-based or markerless tracking allows the detection of real elements such as electronic boards and is more robust than marker-based tracking against partial overlap, i.e. the object to be detected doesn't have to be always in full view of the camera. Thus, this method has been selected as the main tracking system.[1]

[1]The process and parameters of the algorithms described in this section have been adapted during the development stage to achieve the best results from the images of the E2LP electronic boards.

Fig. 4 Partial detection of the board on a close-up. MMC component highlighted in AR

Image-based tracking is performed by searching characteristic points (key points) or features of the images, using in this case corner-based feature detector algorithms. In this area, OpenCV offers a set of algorithms included in the package Features2D.

The main algorithm selected for this task has been ORB (Oriented BRIEF) [11]. This algorithm is implemented in OpenCV and has the characteristic of being invariant to image rotation offering detection of partially rotated pictures. This feature has been considered as essential for the project because it erases possible problems deriving from the fact of not knowing the initial position of the board regarding the camera.

For the tracking process, **two different datasets** (groups of pictures to be detected) have been defined: one containing the image of the whole board and another one containing the four main quadrants and two of the lateral views of the board.

The reason of this classification is the use of the magnifying glass: when users are pointing at the whole board only one image of it is required. However, if they want a closer look, only part of the board will be in view requiring a smaller section of the board to be compared with (Fig. 4).

Once extracted the key points of the image, BRISK (Binary Robust Invariant Scalable Keypoints) [12] is used to extract the **descriptor** vector.

The process of finding frame-to-frame correspondences can be formulated as the search of the nearest neighbour from one set of descriptors for every element of another set. It's called the "matching" procedure. There are two main algorithms for descriptor matching in OpenCV: **Brute-force matcher** and FLANN-based matcher. In this case, the former has been due to its better permanence with ORB.

To improve the results (remove false matches) the **KNN** (K Nearest Neighbour) algorithm is used, that determines the probability of a detected point to be correct based on its surrounding points and then RANSAC (Random Sample Consensus).

Figures 5 and 6 present the matches calculated without KNN and with KNN filtering respectively. Figure 7 illustrates the improvement after RANSAC is used.

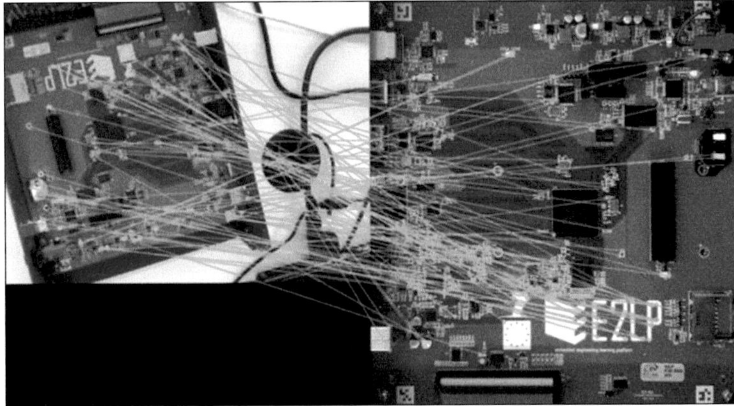

Fig. 5 Matches without KNN filtering. Image from the webcam (*left*) and saved original (*right*)

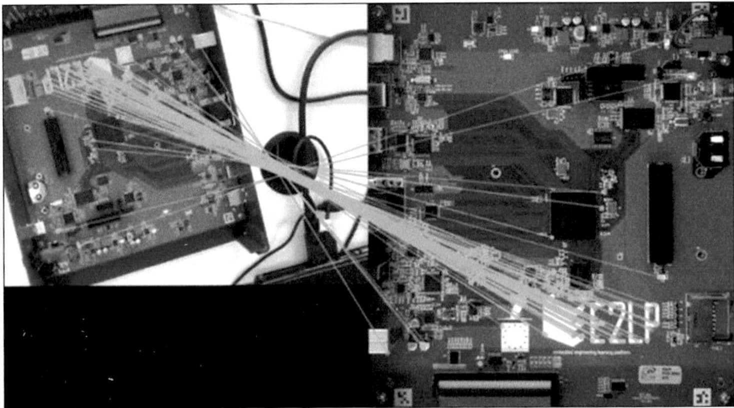

Fig. 6 Matches with KNN filtering. Image from the webcam (*left*) and saved original (*right*)

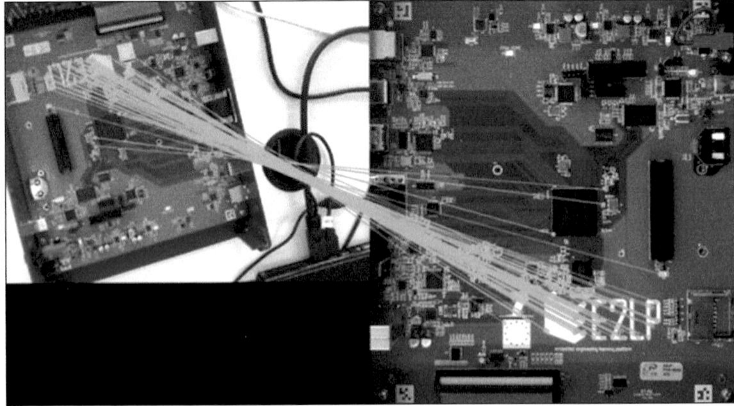

Fig. 7 Matches after RANSAC algorithm. Image from the webcam (*left*) and saved original (*right*)

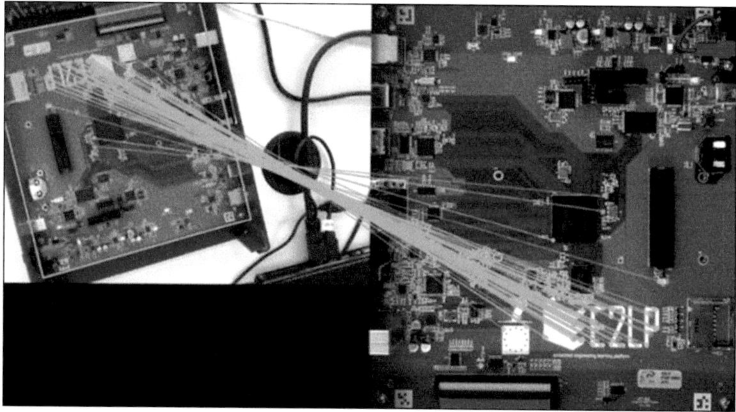

Fig. 8 Final matches after the second processing. Image from the webcam (*left*) and saved original (*right*)

In order to improve the stability of the system and obtain a better superposition of the AR elements, another stage has been added to the processing system. This new stage refines the homography obtained from the previous stage after generating a new matching process but with the frame coming from the camera rotated, obtaining the results shown in Fig. 8.

Although this process adds computational load to the overall system it provides a greater stability and accuracy in the calculations of the pose matrix estimation for the AR components, avoiding fluctuations and fixing them.

4.1.2 Marker-Based Tracking

This kind of tracking is based on the recognition of some patterns (called markers) with very specific characteristics (see Fig. 3): they are square and black and white (or two colours with big contrast between them).

In this project the marker-based tracking has been implemented as a secondary tracking system to be used in specific cases where the image based recognition system is not fast or reliable enough, and as a secondary calibration method.

The software detection of these markers is accomplished as follows:

- Conversion to greyscale of the frames coming from the camera.
- Binarization (i.e. black and white conversion) according to a threshold.
- Detection in the resulting image of 1–4 markers at the same time.
- Removal of erroneous matches.
- Decodification.
- Calculation of the rotation/translation matrixes of the markers.

4.2 Users' Interfaces

Taking into account the objectives set at the beginning of the document, the user interface has been developed following mobile devices' designs. It consists of a main window where augmented reality is displayed, a lateral bar on the left side for navigation purposes and a message bar on top of the main window to display information and instructions. When users first execute the software they access directly the "Board Discovery Mode". In this mode, when they point at the electronic board with the camera they can see the names of the main components on top of them (Fig. 9).

When touching a component (either clicking with the mouse if using a regular screen or with the finger if using a touchscreen) the datasheet of the component is automatically loaded on the screen. This way, students don't need to go through books or PDFs to locate its technical specifications.

For the beta software five exercises from three different subjects of Computer Engineering have been chosen and developed and are currently being within the universities of the consortium.

Attending to the classification in Table 1, each exercise provides, apart from regular theory and requirements PDFs, augmented information such as the components to be used during the exercise (those components appear highlighted, similar to Fig. 9 and when touching them an explanation of their functioning is displayed) or instructions to follow (step by step indication of which interfaces and components have to be connected). In addition, an extra exercise, completely different from the others and called "E2LP Board Discovery Exercise" has also been codified. This exercise serves as the first contact with the E2LP main board: through three levels of increasing difficulty, students have to demonstrate their knowledge of the main components of the board by pressing them when asked.

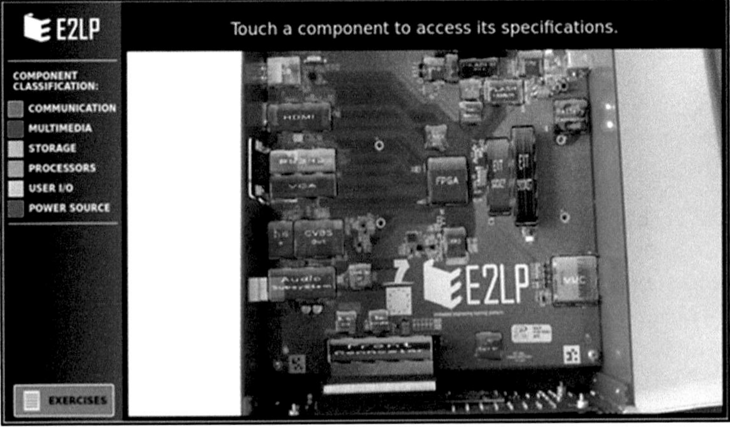

Fig. 9 User interface in board discovery mode

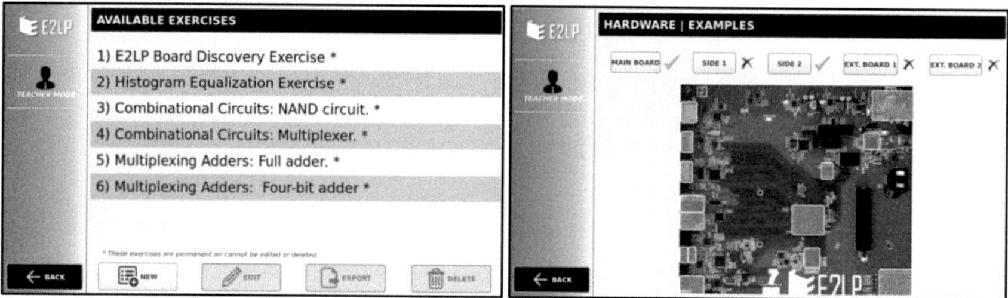

Fig. 10 Teacher interface. General window (*left*) AR display option (*right*)

4.2.1 Teacher Interface

Parallel to the user interface, a new one called "Teacher Interface" has been designed and developed, to allow educators to add their own exercises with AR features to the system.

The interface has been designed following the same objectives of usability of the user interface, in order to avoid a long learning process. Accessible through the user interface, it allows the creation, edition, deletion and import/exportation of exercises (Fig. 10). The last characteristic is though for an easy exchange of exercises between different universities or centres.

5 ARI Components: Hardware

In order to make the ARI fully interactive with users, its hardware components (articulated arm's tracking and pointer's vibrotactile systems) must be connected to the main software system and interface.

5.1 Articulated Arm

The AR software is able to communicate with the arm through a serial protocol that allows the software to erase a certain level of uncertainty making use of the geometric model (Fig. 11) of the arm to delimit the group of images to be compared. In this way, the process is accelerated and the general system is more robust.

In addition, once the board has been detected, the system stops processing frames, meaning that it saves resources for other tasks and allows also users to interact with the board (e.g. connecting cables or using the tactile pointer) without losing the AR elements from the screen (Fig. 12).

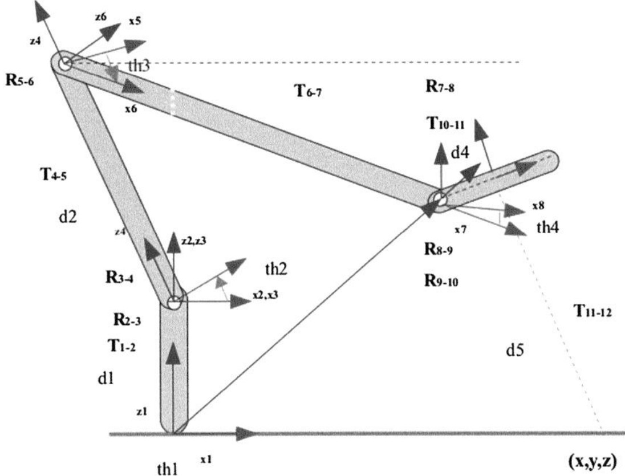

Fig. 11 Geometric model of the articulated arm

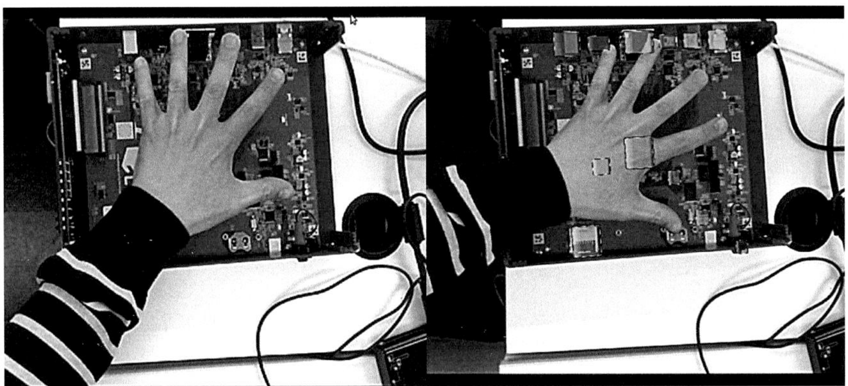

Fig. 12 Improvement of stability of the AR system by adding the use of the arm

To accomplish this purpose, the following hardware development process has been followed:

Once the mathematical model of the arm has been identified, a sensing strategy has been implemented. The criterion of choice is a compromise between the less expensive solution, the most robust and the easiest to integrate without modifying the mechanical structure. The best identified compromise is the use of MEMS inclinometers sensors. However, because inclinometer measures the angle with respect to the gravity axis, it means the angular position of an axis collinear to the gravity such as the vertical articulation of the arm should be measured with another sensor.

To solve this limitation, a specific absolute magnetic angular sensor has been implemented as showed on Figs. 13, 14 and 15. The specific PCB developed allows the integration of the sensor inside the base of the articulated arm but also to collect data from inclinometer using SPI BUS and a low cost 8bit Microcontroller.

Fig. 13 Magnet fixed on a spring

Fig. 14 Sensor working principle

Fig. 15 Integration inside the arm base

5.2 *Tactile Pointer*

The tactile pointer acts as component selector, allowing users to see the datasheet information of the selected component as they would do through the touchscreen (Fig. 16).

The technical principle of this approach consists in generating a magnetic field under points of interest of the E2LP board and detects them thanks to a magnetometer embedded inside the interactive pen as shown on Fig. 17.

Figure 18 shows that the MAG-ID board is placed under the E2LP board inside the blue box and shows also a zoom-in picture of one inductor printed on the 2 layers MAG-ID PCB allowing a generation of constant magnetic field.

Fig. 16 Some of the components are accessible with both the touchscreen and pointer

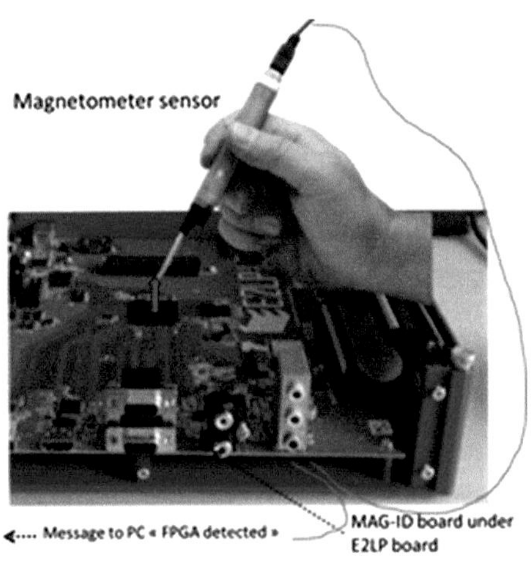

Fig. 17 Working principle

Fig. 18 The "MAG-ID" detection board

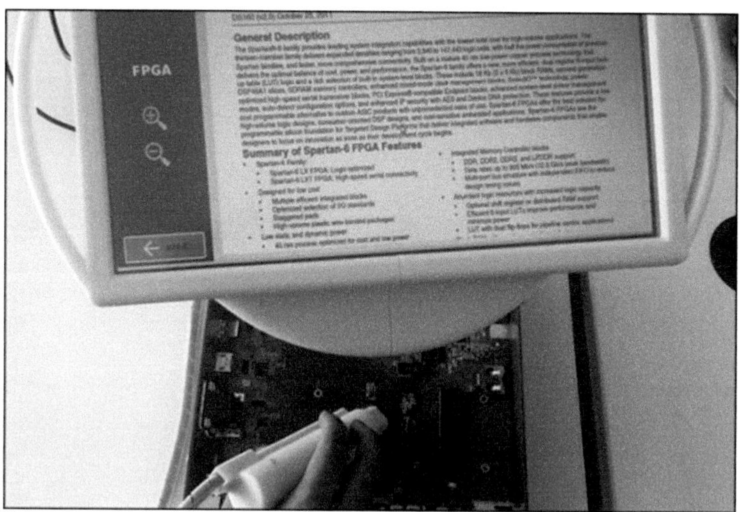

Fig. 19 Tactile pointer launching information of the touched component

The "MAG-ID PCB" is composed of 31 printed inductors or "magnetic tag" with different shapes depending on the size of the targeted component.

The dissipated power reaches 0.75 W allowing the MAG-ID board to be USB self-powered compliant. However, this approach implies the magnetic tags should be switched-on one by one and lies to a magnetometer data acquisition at each magnetic tag commutation. Considering a magnetometer refresh rate of 160 Hz, to check the 31 magnetic tags, the overall refresh rate of the tracking system is lower than 5 Hz, which is acceptable for a human machine pointing application.

Similar to the articulated arm, the AR software receives through USB signals the component chosen with tactile pointer and displays the information on the screen (Fig. 19).

5.3 *Tactile Feedback Accessory*

Besides supplying the localization of the zone of interest for the augmented reality process, the role of the Haptic pen is to "make tangible the invisible" thanks to innovative interactions metaphors. The objective of this system is to allow the student "to feel" the physicals characteristics that describe an electronic circuit as the frequency, the nature (analogical or digital) or the flowing current intensity.

This approach, based on a tangible object, has to incite the student to explore an invisible world and so to become aware by the experiment of the function of the various components which compose an electronics board. To carry out this function, the same pen than used for localization purposes is used and includes 3 different vibration range electro-magnetic actuators (Fig. 20).

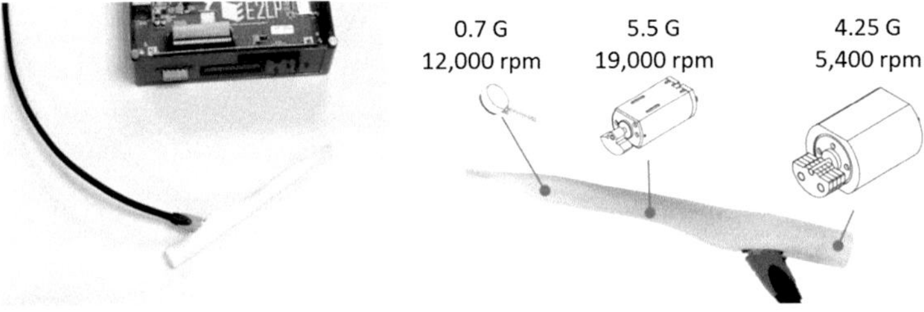

Fig. 20 Actuators inside the tactile pen

Vibrotactile classification metaphors	Symbolic definition
Analog+HF+Continuous	Electronic components dedicated to high frequency analogic signals working continuously
Analog+LF+Continuous	Electronic components dedicated to low frequency analogic signals working continuously
"Clac Clac"	Electronic components dedicated to one manual action (power on, jumper setting)
Digital+HF+Continuous	Electronic components dedicated to high frequency digital signals working continuously
Digital+HF+Discontinuous	Electronic components dedicated to high frequency digital signals working in firmware context (Controller, memory, I/O…)
Hybrid Digital-Analog Continuous	Electronic components dedicated to high frequency digital & analog signals supposed working continuously (Video & audio analogic to digital converters)
Hybrid Digital-Analog Discontinuous	Electronic components dedicated to high frequency digital & analog signals supposed working in firmware context (Mixed signal microcontroller, connectors with others boards)

Fig. 21 Classification used for the tactile metaphors implementation

Figure 21 shows 7 classifications used to develop tactile metaphors. The motivation of this classification is to help students memorize thanks to the sense of touch.

This classification corresponds to the different families of function that we can find on a typical electronic board. The objective of the tactile accessory is now to create 7 different vibrotactile metaphors patterns easy to distinguish without ambiguity.

Figure 22 shows an example of one vibrotactile pattern implementation.

It appears that the vibrotactile feedback is well perceived by the user thanks to the wide range of frequency and amplitude vibration provided by the innovative combination of vibrotactile actuators. The next important step is the user evaluation to check if metaphors are relevant and effectively help students to memorize electronics concepts easier than before.

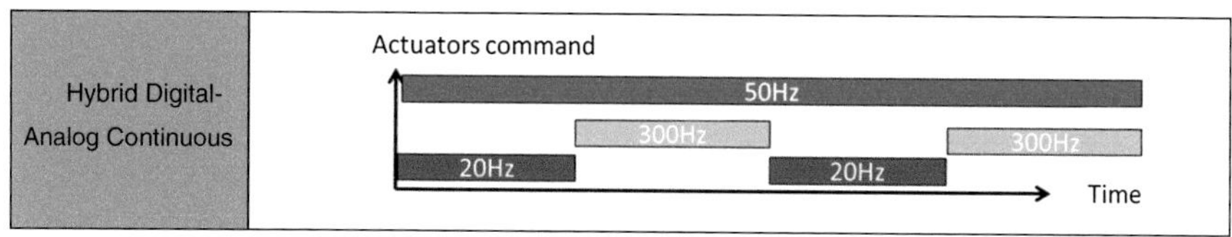

Fig. 22 Example of metaphor with three actuators

6 Conclusions and Scalability of ARI

The augmented reality interface described in this document consists of an interactive and user-friendly system aimed at helping students discover and use the E2LP boards through augmented reality and vibrotactile capabilities.

At the present moment, apart from the E2LP main board, the ARI also includes the detection of two E2LP extension boards.

However, the versatility of the ARI's tracking system would allow the inclusion of new boards developed outside the scope of this project. This capability supports the growth of the E2LP system once the project finishes, opening also the door at the possibility of students creating their own extension boards and presenting them to the rest of the class or the teacher using the AR interface, as a new and more collaborative and interactive way of learning.

References

1. Fantini, M., Persiani, F., Di Stefano, L., Azzari, P., De Crescenzio, F.: Augmented reality for aircraft maintenance training and operations support. IEEE Comput. Graph. Appl. **31**(1), 96–101 (2011)
2. Henderson, S., Feiner, S.: Exploring the benefits of augmented reality documentation for maintenance and repair. IEEE Trans. Visual. Comput. Graphics **17**(10), 1355–1368 (2011)
3. Matsutomo, S., Miyauchi, T., Noguchi, S., Yamashita, H.: Real-time visualization system of magnetic field utilizing augmented reality technology for education. IEEE Trans. Magn. **48**(2) (2012)
4. Billinghurst, M., Dünser, A.: Augmented reality in the classroom. Computer **45**(7), 56–63 (2012)
5. Temerinac, M., Kastelan, I., Skala, K., Rogina, B.M., Reindl, L., Souvestre, F., et al.: E2LP: a unified embedded engineering learning platform. In: 16th Euromicro Conference on Digital System Design DSD 2013, Special session on European projects on digital systems design, Santander, Spain, 04–06 Sept 2013
6. Anastassova, M., Souvestre, F., Artetxe González, E., Setién Gutiérrez, A., López Benito, J.R., Barak, M.: Learner-centered evaluation of an augmented reality system for embedded engineering education. In: Federated Conference on Computer Science and Information Systems (FedCSIS), Warsaw, Poland (2014)
7. Kastelan, I., Barak, M., Sruk, V., Anastassova, M., Temerinac, M.: An approach to the evaluation of embedded engineering study programs. In: 36th International Convention on Information and Communication Technology, Electronics and Microelectronics MIPRO, Opatija, Croatia (2013)
8. OpenCV. http://opencv.org/ (2015). Accessed 20 July 2015

9. Pulli, K., Baksheev, A., Kornyakov, K., Eruhimov, V.: Realtime computer vision with OpenCV. http://queue.acm.org/detail.cfm?id=2206309 (2012). Accessed 22 April 2012
10. OpenGL. http://www.opengl.org/ (2015). Accessed 20 July 2015
11. Rublee, E., Rabaud, V., Konolige, K., Bradski, G.R.: ORB: an efficient alternative to SIFT or SURF. In: ICCV 2011, pp. 2564–2571
12. Leutenegger, S., Chli, M., Siegwart, R.: BRISK: binary robust invariant scalable keypoints. In: ICCV 2011, pp. 2548–2555

E2LP Remote Laboratory: e-Learning Service for Embedded Systems Education

Rafał Kłoda and Jan Piwiński

Abstract Embedded Engineering Learning Platform (E2LP) FP7 project deals with Embedded Systems Education at University level, which requires a multi-disciplinary approach, involving different technologies and system solution optimizations. The paper presents the results of Remote Laboratory (RL) services development for distance learning for Embedded Systems Education, developed under E2LP. The paper addresses advanced information technologies solutions in the integration stages along with novel hardware technologies involved. E2LP RL assembles hardware and the exercise materials provided by the project partners and it delivers secure and open access e-learning portal, which allows to create full course and provide alternative teaching methods through the real-time experiments. This paper could be also a guideline for future RL developers in embedded systems domain. We describe what components are needed to build remote laboratories for embedded systems and what could be a scope of remote operation for users.

Keywords E2LP · Remote laboratory · Curriculum integration · Embedded systems

1 Introduction

As embedded software systems have grown in number, complexity, and importance in the modern world, a corresponding need to teach computer science students how to effectively engineer such systems has arisen [1].

R. Kłoda (✉) · J. Piwiński
Industrial Research Institute for Automation and Measurements PIAP,
Al. Jerozolimskie 202, 02-486, Warsaw, Poland
e-mail: rkloda@piap.pl

J. Piwiński
e-mail: jpiwinski@piap.pl

R. Szewczyk et al. (eds.), *Embedded Engineering Education*,
Advances in Intelligent Systems and Computing 421,
DOI 10.1007/978-3-319-27540-6_7

While embedded systems comprise about 99 % of the entire computer market [2], many undergraduate computer engineering programs still teach programming and design skills that are applicable to a general-purpose computer rather than to the more specialized embedded systems [3].

Early exposure to embedded computing systems is crucial for students to be prepared for the embedded computing demands of today's world. However, exposure to systems knowledge often comes too late in the curriculum to stimulate students' interests and to provide a meaningful difference in how they direct their choice of electives for future education and careers [4].

Those aforementioned issues were a genesis to create an unified learning platform, customized to embedded systems curriculum and was the main goal of the E2LP project.

Embedded Computer Engineering Learning Platform (E2LP) is a European FP7 project of 3 years duration, started in September 2012 [5].

In E2LP project a Remote Laboratory is an experiment, demonstration and a process running locally to design and control an experiment board based on a FPGA device, but with the ability to be monitored and controlled over the Internet (E-learning portal).

In the base case, the RL can be an experiment board connected to a computer through a standard interface and with the host computer connected to the Internet, which provides a remote access. The client can be any computer connected to the Internet with an ability to see the same interface as the local host as well as has the same programs, interfaces and modules.

RL framework consists of three main elements:

1. E-learning portal. This part of RL provides an access to knowledge (on-line exercises, data sheets) as well as remote operations with E2LP main board through a web user interfaces.
2. Laboratory hardware. Main element is E2LP experimental board with programming cable device and other equipment to conduct remote learning process (E2LP server, digital card, serial port server).
3. Laboratory software. It includes the necessary software to programming board and other applications/services/interfaces based on several IT technologies, which provide proper functioning of the whole Remote Laboratory and their hardware components. Here there are also a number of communication ports, which provide flawless operation of specific applications and services in E2LP server as well as in several cases enable user to individually configure the communication with a given device.

2 Remote Laboratories in Learning Courses

As technology is increasingly being seen as a facilitator to learning, open remote laboratories are increasingly available and in widespread use around the world [6].

Virtual and remote laboratories (VRLs) are e-learning resources that enhance the accessibility of experimental setups providing a distance teaching framework which meets the student's hands-on learning needs [7]. They have been considered as one of the five major shifts in a century of engineering education, thanks to the influence of information and computational technologies [8].

An important study of the implementation of VRLs into learning courses was reported here [9]. This study presents the results of integrating the open remote laboratories into several courses, in various contexts and using various methodologies. These integrations, all related to higher education engineering, were designed by teachers with different perspectives to achieve a range of learning outcomes.

RL have been widely popular among many universities. They are built in order to enhance learning and minimize the gap between theory and practice. Remote laboratories provide on-line pervasive workbenches, which allow an interactive learning environment that maintains student attention.

Laboratories, which are found in all engineering and science programs, are an essential part of the education experience. Not only do laboratories demonstrate course concepts and ideas, but they also bring the course theory into practice. In a traditional laboratory, the user interacts directly with the equipment by performing physical actions (e.g. manipulating with the hands, pressing buttons, turning knobs) and receiving sensory feedback (visual and audio). However, equipping a laboratory is a major expense and its maintenance can be difficult [10].

Since the experiments are performed in a laboratory that contains expensive equipment, the students must be supervised which limits the time they have. This also requires a class with many groups performing the experiment at the same time, and thus many instruments are required to support each group. Laboratory experiments are also a serious problem for distance learning students who may not have an access to the laboratory at all [11].

However a review of literature highlighted the lack of meaningful assessment tools for virtual laboratory environments in engineering education. Literature review also highlighted the shortcomings of traditional student lab work assessment practices, among engineering faculties. There are also the problems of time demands, bias, inconsistency, and increasing student numbers which make the traditional lab work assessment scheme impractical [12].

VRLs should focused to support a specific group of students, those who are marginalized by not being able to regularly attend lessons and perform the laboratory exercises as well as should be facilitated to learning process to those students. Teachers and institutions should not just provide open access to their educational material, but should openly share their experiences, case studies, lessons learned, and suggestions to improve the teaching and learning processes. Although busy students are required to demonstrate their final projects on the actual hardware, the Massive Open Online Course (MOOC) platforms enable them to prepare for this at home, and then to be able to demonstrate valid hardware results in the laboratory [13].

It should be stressed that Remote Laboratories cannot replace the classical education course. There are of course drawbacks of implementation such tools, mainly in lack of communication between student and course supervisor. This type of systems can isolate students and reduce their motivation in learning process. Furthermore, students could not receive instant feedback from their questions and cannot talk in real-time about results obtained in the learning activities with the teacher.

3 E2LP RL Concept of Design and Project Research Objectives

In E2LP project a Remote Laboratory is a service, which enable students to access the laboratory equipment and execute remote operations to carry out exercises. The main goal of RL was implementation of instant feedback from remote E2LP board in a way that user would operate with the real board as if it was connected locally. This functionality was a purpose to develop the GUI web interface of E2LP board front panel that exactly reflects the real board, which has connections to real signals from the real board.

The main advantage of proposed E2LP RL framework (Fig. 1) is a possibility for students to interact with the real E2LP platform interfaces, implemented as a web services in Moodle [14] and work with software applications, on the same operational level like they are actually operating the same tools and instruments in classic lesson in laboratory.

RL is a gate which provides an access to continuously refreshed interfaces and signals from the real board and enable users to remotely control and program the board directly from their computer at home, having instant visual feedback.

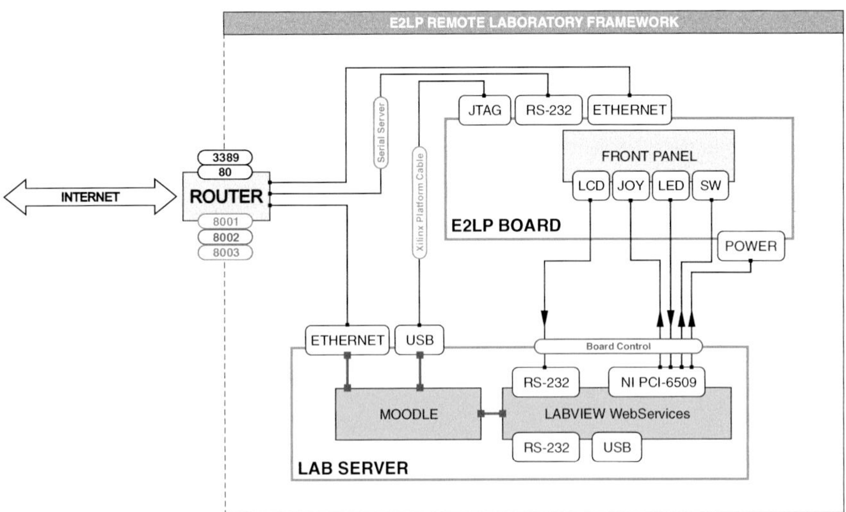

Fig. 1 E2LP remote laboratory framework

To achieve this, it is necessary to forward data directly to the server over common interfaces or over local network by using dedicated hardware solutions and specified proper router configuration.

The E2LP RL should allow users to do following actions over an Internet connection, which are the list of E2LP Remote Laboratory main functionalities:

1. Dedicated software and hardware solutions provide an access to laboratory equipment and enable students to set them up and operate them at the required level to carry out selected exercises.
2. Users could access the essential data sheets, tutorials and software tools, which are available on the E-learning portal as an introduction to the course. Each laboratory exercise is presented in transparent form to the user through tabs and such division is implemented into Moodle based platform for e-learning course (Basic information, Theoretical explanations, Instructions, Configure Platform, Feedback, Discussion on results questionnaire for lab evaluation).
3. After booking in a given time slot users could remotely program given set of exercises over the Internet and simultaneously, in real time, could monitor the evolution of the experiment on implemented dedicated Graphical User interface (GUI) of the Front Panel of the real E2LP board.
4. Automatic verification of course assignments will allow an advanced management of assignments and submissions together with feedback information mechanisms for both teachers and students, which will verify, whether the students designs are correct or not according to the specifications.

4 RL Development and Implementation Phases

Connection with the Remote Laboratory is provided via e-learning portal, which is based on Apache server, PHP and SQL server. It provides an access to knowledge (exercises, data sheets) and laboratory hardware through a web user interfaces. The second role of e-learning portal is management of users, which means enable them access to the laboratory hardware and software (booking functionality and authorization). In E2LP project the e-learning platform is based on Moodle Platform, which is one of the most popular open source learning management systems. The URL of the Remote Laboratory portal is [15]:

RL presents fully operational and tested system, which is enriched with dedicated modules to E2LP Mother Board, which provide real-time remote control, monitoring and programming. Below we show the main advantages of the system:

- The final laboratory exercise on the web has sections (tabs) to enable user to have the full experience of working on the laboratory exercise. These are Digital System Design course exercises, which aim is to control Switches, JOY Push Buttons, LEDs, LCD output in the front panel of the E2LP board as well as RS-232 port are available for remote operations.

- Advance booking system, which enables to reserve a time slot for individual remotely tests of the solution for a given exercise. Booking functionality enables to access up to 4 remote E2LP boards.
- The fast bit file loading module enables remote configuration and immediate respond of the successful E2LP board configuration, without a requirement for users to have a specialized Xilinx [16] software to do it.
- The user friendly Graphical User Interface of the Front Panel, which reflects to the same panel on the real E2LP board, enables user to monitor and control remotely each switch, button, LED and LCD output. The GUI is enriched with the checking correctness of the solution module, which compares the students solution with a master, created by the teacher.
- Automatic verification module, which is based on regular expressions, checks the correctness of the users solution. The pattern for solution is prepared by the teacher or course creator. After comparison the user is informed visually about correctness of his solution.
- The 'Discussion on results' functionality module consist the output information from check correctness solution module, by showing the log records output from the E2LP board Front Panel and enable Teacher and user to exchange information about given exercise.

The whole environment is managed by powerful E2LP Server, controlled by LabVIEW software, which is equipped with all common interfaces, which are essential for internal hardware and software compatibility. E2LP Server is connected via Ethernet interface to the local network, which is responsible for seamless data communication between environment's components. The crucial component of the remotely controlled environment is an experiment base board, which is controlled by programming device (Xilinx Platform HW-USB-II-G). This programming device provides integrated firmware to deliver high-performance, reliable and user-friendly configuration of the base board and enables user to program other Xilinx CPLD devices. This programming device is fully integrated and optimized for use with specialized Xilinx iMPACT software, which enable users to perform remote operations such as programming and configuring FPGA via JTAG interface.

The NI PCI-6509 digital card with 96 bidirectional I/O lines enable user by dedicated GUI interface (Fig. 2) to control each pin in the boards front panel interface and consequently enable him to control each led, switch and button.

Furthermore specific converter communicates with LCD pins on boards front panel interface and translates them into RS232 ASCII chars. This converter was designed to enable remote characters reading from LCD 2X16 chars (Fig. 3).

Converter is connected to PCB controller of LCD (or together with LCD) and reads LCD's control commands as well as its input data. Based on that converter creates in RAM memory the data, which is corresponding to this which should be displayed on LCD. This data we can read via USB interface (in RS232 emulation mode) after a query from our computer. Thank to this we are enabled to send remotely the content of LCD, in limited data transfer (especially in comparison to transferring the image of LCD state in graphic file form).

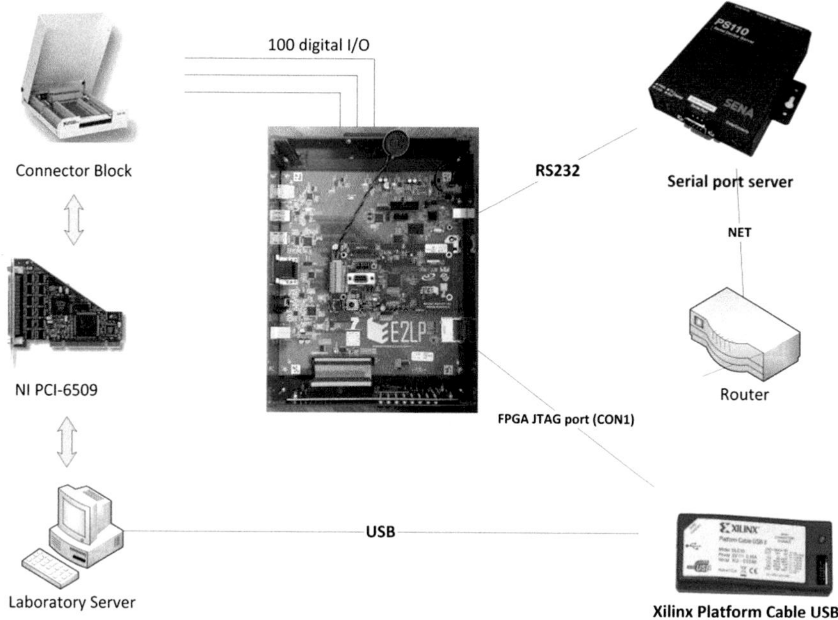

Fig. 2 Remote laboratory concept of solution

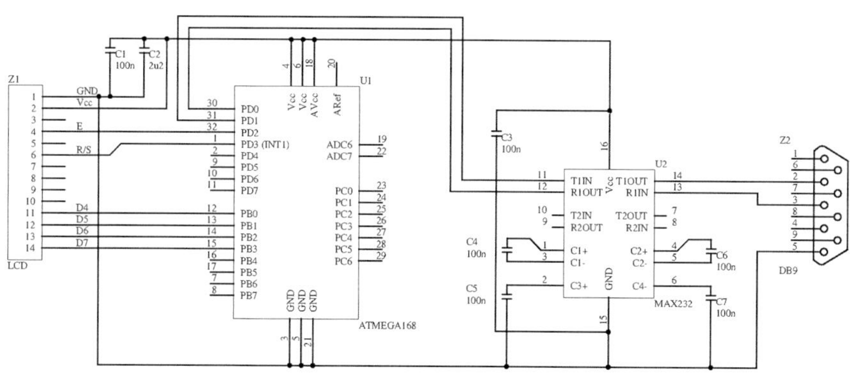

Fig. 3 LCD controlling signal converter to RS232 schema

Converter is built based on ATMEL-ATMega processor where we used 4 lines of port B (data lines of LCD) and 2 lines of port D (INT0, D3). Processor process the input data to LCD in 4 bit form. Data processing takes place in response to falling edge INT0 input signal (LCD signal E). In the first cycle the state of line D3 is read (signal R/S of LCD) and dependently on its level further analysis take place on data or control command.

The converter doesn't operate the commands, which define additional chars (access to chars memory) as well as 8 bit stream format mode. It is impossible to read cursor location and its blinking mode.

The system sends data from LCD after forwarding to it any char over RS-232. In result of acquiring the char, the content of the buffer is sent—two lines of the text with 16 chars, which are finished with CR and LF chars (this is done due to easy read content using the terminal, which operates RS232).

Microcontroller system and level converter (MAX232) were placed on small PCB with cable sockets, which enable to connect to GOLDPIN joint.

In order to control the physical signals on E2LP Board the NI 6509 device was used as a low-cost solution with advanced features. This device offer 96 digital I/O lines (5 V TTL/CMOS), high-current drive (24 mA sink or source) and the most important features such programmable power-up states. Each 8-bit port on an NI 6509 can be input or output as well as in this device it is possible to configure digital lines as high-impedance input, high output, or low output. The programmable power-up states provides, that the outputs never go through an incorrect state during power up. This device is used to is control all switches and all buttons and monitors each LED on E2LP Board as well as it is possible to provide access up to four E2LP Boards for remote operation.

To integrate physical layer (NI 6509 device) with application layer (user interface) the Web Service (WS) (Fig. 4) was developed using LabVIEW environment. This Web Service has its own user interface (web application) that provide access to control and monitor physical signals on E2LP Board. This web application uses AJAX programming technique that quickly respond to user requests. AJAX enables JavaScript to communicate directly with the Web Service using the XMLHttpRequest object, which requests and updates only the required data instead of reloading whole interface. This concept also enables easy integration with any e-learning systems [17].

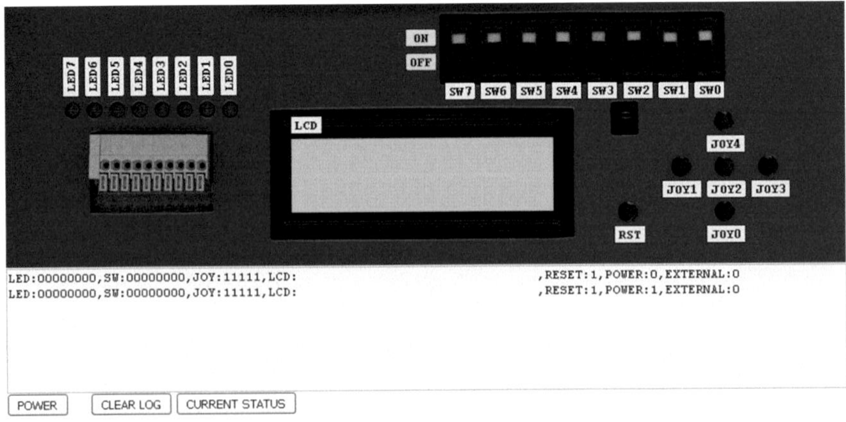

Fig. 4 E2LP board front panel web service

5 Conclusions

The new systems do not seek to replace teachers, but rather to assist them by releasing them from monitoring the use that students make of the Internet and by making them more available to enhance the learning experience [18].

This paper has discussed all the features provided by E2LP RL for its implementation and deployment in embedded systems engineering education along with feedback from the universities that had deployed it in their learning curricula. Robust RL portal enables users to access E2LP platform over the Internet, configure it compiling VHDL code and having the immediate feedback of solution on their own computer.

Results presented in the paper confirms that introduction of RL into curriculum and new learning model is challenging in the education of engineers in embedded systems. Proposed solutions based on integrated together Remote Laboratory components and e-learning Moodle Platform enable student to acquire desired knowledge about digital systems and significantly support learning process.

Acknowledgments E2LP Remote Laboratory development were performed in the Industrial Research Institute for Automation and Measurements PIAP.

References

1. Mattmann, C.A., Medvidović, N., Malek, S., Edwards, G., Banerjee, S.: A middleware platform for providing mobile and embedded computing instruction to software engineering students. IEEE Trans. Educ. **55**(3), 425 (2012)
2. Ganssle, J.: Embedded Y2K. Embedded Syst. Program. **3**(17), 97–99 (1999)
3. Jackson, D.J., Caspi, P.: Embedded systems education: future directions, initiatives, and cooperation. SIGBED Rev. **2**(4), 1–4 (2005) (Discussion)
4. Benson, B., Arfaee, A., Kim, C., Kastner, R., Gupta, R.K.: Integrating embedded computing systems into high school and early undergraduate education. IEEE Trans. Educ. **54**(2), 197 (2011)
5. E2LP Project Website. http://www.e2lp.org
6. Marques, M.A., Viegas, M.C., Costa-Lobo, M.C., Fidalgo, A.V., Alves, G.R., Rocha, J.S., Gustavsson, I.: How remote labs impact on course outcomes: various practices using VISIR. IEEE Trans. Educ. **57**(3), 151 (2014)
7. de la Torre, L., Heradio, R., Jara, C.A., Sanchez, J., Dormido, S., Torres, F., Candelas, F.A.: Providing collaborative support to virtual and remote laboratories. IEEE Trans. Learn. Technol. **6**(4), 312 (2013)
8. Froyd, J., Wankat, P., Smith, K.: Five major shifts in 100 years of engineering education. In: Proc. IEEE **100**(Special Centennial Issue) 1344–1360 (2012)
9. Zubía, J.G., Alves, G.R. (eds.): Using Remote Labs in Education. Two Little Ducks in Remote Experimentation. Prize for Best Research UD—Grupo Santander, University of Deusto (2011)
10. Distance-Learning Remote Laboratories using LabVIEW. http://www.ni.com/white-paper/3301/en/. Accessed 06 Sept 2006
11. Nafalski, A., Machotka, J., Nedic, Z.: Collaborative Remote Laboratory NetLab for Experiments in Electrical Engineering in Using Remote Labs in Education. Two Little Ducks in Remote Experimentation, University of Deusto, pp. 177–199 (2011)

12. Achumba, I.E., Azzi, D., Dunn, V.L., Chukwudebe, G.A.: Intelligent performance assessment of students' laboratory work in a virtual electronic laboratory environment. IEEE Trans. Learn. Technol. **6**(2), 103–106 (2013)
13. Ackovska, N., Ristov, S.: OER approach for specific student groups in hardware-based courses. IEEE Trans. Educ. **57**(4), 242–247 (2014)
14. https://moodle.org/
15. E2LP Remote Laboratory, e-learning portal for E2LP. http://e2lp.piap.pl
16. www.xilinx.com/
17. Tawfik, M., Sancristobal, E., Martin, S., Diaz, G., Castro, M.: State-of-the-art remote laboratories for industrial electronics applications. In: Technologies Applied to Electronics Teaching (TAEE), pp. 359–364 (2012)
18. Sanders, D.A., Bergasa-Suso, J.: Inferring learning style from the way students interact with a computer user interface and the WWW. IEEE Trans. Educ. **53**(4), 613–620 (2010)

Advanced Projects and Applications for Embedded Systems Engineering on E2LP Platform

Dario Grgić, Sebastian Böttcher, Marc Pfeifer, Johannes Scherle, Benjamin Völker, Jan Burchard, Sebastian Sester and Leonhard M. Reindl

Abstract The E2LP-Platform is capable to impart many different fields of learning content. Starting from simple, short exercises also complex projects can be realized covering up to several weeks of workload. This work presents a documentation of four students projects developed and performed at University of Freiburg in the Advanced Embedded System Laboratory. This laboratory contains a dynamic classroom approach in which the required laboratory hardware is mobile. Exploring real-wold challenges and problems motivates the participants to acquire a deeper knowledge.

Keywords E2LP · VHDL · Project based learning

D. Grgić (✉) · S. Böttcher · M. Pfeifer · J. Scherle · B. Völker · J. Burchard · S. Sester · L.M. Reindl
Department of Microsystems Engineering—IMTEK, Laboratory for Electrical Instrumentation, University of Freiburg, Georges-Köhler-Allee 106, 79110 Freiburg, Germany
e-mail: grgic@imtek.uni-freiburg.de

S. Böttcher
e-mail: boettchs@informatik.uni-freiburg.de

M. Pfeifer
e-mail: pfeiferm@informatik.uni-freiburg.de

J. Scherle
e-mail: johannes.scherle@gmail.com

B. Völker
e-mail: voelkerb@informatik.uni-freiburg.de

J. Burchard
e-mail: burchard@informatik.uni-freiburg.de

S. Sester
e-mail: sesters@informatik.uni-freiburg.de

L.M. Reindl
e-mail: reindl@imtek.uni-freiburg.de

R. Szewczyk et al. (eds.), *Embedded Engineering Education*, Advances in Intelligent Systems and Computing 421,
DOI 10.1007/978-3-319-27540-6_8

1 Audio Codec Implementation in VHDL

1.1 Project Introduction

In this Advanced Embedded Systems Lab (AESL) project, the onboard audio codec is used to play PCM audio data from a WAV file via the E2LP-board. The goal of the project is to read in data with the SD card module, store the file in DDR2 or VHDL synthesized block RAM and proceed the data via the audio chip. Additionally functions to control the playback can be programmed and customized.

Current version of the project uses block RAM to store audio data which is sent via the Armada extension board. This data is then continuously sent to the audio chip for playback. Additionally, a volume control module is implemented, and the audio chip can be switched into bypass mode (Line-In is directly connected to the headphones).

1.2 VHDL Modules Overview

The VHDL code consists of several different modules, which are briefly described here. Further information on single aspects of the modules can be found in the documentation.

initModule: This module initializes the audio chip. It powers the chip on, enables SPI configuration, and then uses the spiModule to configure the chips registers. It uses a state machine to properly go through the different registers/configuration stages. For more information see Sect. 1.3.

spiModule: Takes a 32-bit logic vector to send to the chip via the SPI protocol.

volModule: Two of the boards push buttons are used here to change the chips internal volume settings. Whenever a button is pushed, the spiModule is used to reconfigure the relevant chip registers to a new volume level. Additionally some of the boards LEDs are used to indicate the current volume level.

I²SModule: This is the main module of the program. Here, the block RAM core (see Sect. 1.5) is used to store data received via I²C from the Armada board and is sent to the audio chip using the chip-generated LRCLK and BCLK. See Sect. 1.4 for more information.

1.3 Audio Chip Configuration

To use the ADAU1772 audio codec [1] it is necessary to properly configure the hardware to individual requirements. The chip has several different registers that store the current configuration. These registers can be set via I²C or SPI. Initially, the chip is in I²C mode. Since the configuration is done via SPI, the first thing to do is to

Table 1 Registers and their configurations used in project

Register	3rd Byte	4th Byte	Description
0x0000	0000 0000	0000 0111	Main clock en, clock div
0x0008	0000 1000	0000 0001	Regulator active, 1.1 V
0x0009	0000 1001	0000 0101	Core on, bank A, 192 kHz
0x000B	0000 1011	0000 0011	Limiter en, core clock en
0x001B	0001 1011	0000 0001	Unmute both ADC, 192 kHz
0x001D	0001 1101	0000 0011	HP off, polarity, source en ADCs
0x001F	0001 1111	0011 0110	Volume = −20 dB (0.1)
0x0020	0010 0000	0011 0110	Volume = −20 dB (0.1)
0x0023	0010 0011	0001 0000	PGA off
0x0024	0010 0100	0001 0000	PGA off
0x002A	0010 1010	0000 0011	ACD0/ACD1 to DAC0/DAC1
0x002E	0010 1110	0000 0011	Polarity normal, Unmute, enable
0x002F	0010 1111	0011 0110	Volume = −20 dB (0.1)
0x0030	0011 0000	0011 0110	Volume = −20 dB (0.1)
0x0031	0011 0001	0000 0000	Headphone mute disable
0x0032	0011 0010	0000 0001	Sampling rate 8 kHz
0x0033	0011 0011	0000 0001	I^2S master, 32 BCLKs/channel
0x003C	0011 1100	0000 1111	DSP-Bypass on MP4
0x0043	0100 0011	0011 0000	HP output mode, en output stages
0x0044	0100 0100	0000 0011	ADC0/1 filter on

switch the chip into SPI mode. After powering on the chip by pulling PD to 0 (low), SPI mode can be enabled by pulling SS to 0 (low) three times. In this project this is done manually, but can also be achieved by simply issuing three consecutive SPI writes (which are ignored). The chip can now be configured using the standard SPI-protocol. Each SPI transmission sends 4 bytes to the chip: the first byte indicates a read or write action, the second and third byte contain the register address, and the fourth byte contains the actual configuration that the indicated register is to be set to. It is also possible to read in the current configuration of a register, but in this project I only write to the registers, so R/W bit in the first byte is always set to 0. Table 1 shows the registers that are set in the project. The first and second Bytes are all zero, but cannot be omitted during transmission. A short description of each configuration is given, and more extensive explanations of the possible settings of each register and the SPI process can be found in the ADAU1772 manual [1].

One more thing to note is that for enabling the headphone output, first register no. 0x43 has to be set to enable the output stages. Then, to avoid any pops when enabling the headphones, there is a wait period of at least 6 ms until the headphones can be unmuted by setting the register no. 0x31. This wait period allows an internal capacitor to precharge before it is used.

After setting the chip to this configuration, the setup can be tested by triggering the MP4 pin (see E2LP-board documentation), e.g. with one of the push buttons. When MP4 is enabled, the DSP-Bypass is enabled (see reg. 0x3C), which directly puts the Line-In through to the headphone jack. So to test, one can connect some music player to the Line-In jack (top) and some headphones to the headphone jack (middle), and if audio playback is given whenever the MP4 pin is enabled the configuration works properly.

1.4 I^2S Format and Transmission

The I^2S Format (Integrated Interchip Sound) uses 3 lines to transmit audio data between chips. The LRCLK line (left/right clock) is the main clock, indicating the current word and channel. During its low phase, the left data sample is transmitted, and during its high phase the right data sample is sent. It should be as fast as the sampling rate of the audio data that is transmitted. The BCLK line (bit clock) indicates the current bit of the current sample to be sent. It is usually 32 times faster than LRCLK, to allow transmission of audio data with a maximum of 32-bit sample width. The SDATA line is the data line that is usually changed on a falling edge of BCLK. Most of these properties can be changed in the chips configuration registers (see also Sect. 1.3).

After configuring the chip to be the I^2S master in the previous section, it sends the configured clocks to the LRCLK and BCLK pins. Both are used to properly send the audio data to the chip via the SDATA pin. The configuration sets the LRCLK frequency to 8 kHz, which is the lowest possible. I chose this to be able to play as much audio as possible. With the single block RAM that is currently used that is about 16 s of 8 kHz, 16-bit mono PCM data. The left channel data is sent on a low LRCLK, the right channel on a high LRCLK. With each falling edge of the LRCLK the current sample is changed to the next one. Then one BCLK cycle is skipped and on the second BCLK falling edge, the first bit of the current 16-bit sample is put on the SDATA pin. After 16 BCLK cycles, SDATA is again pulled low until the rising edge of LRCLK occurs. This indicates that the right channel data is now to be sent. In this project mono is used to double the amount of data that can be played, so the sample stays the same. Again, all 16 bits of the current sample are sent after one skipped BCLK cycle, just like for the left channel. This process is repeated for all stored samples, until the last sample has been sent, upon which the current sample loops back to the first stored sample.

1.5 Block RAM Core

All data is stored in a single IP block RAM core, which is generated using the *Xilinx LogiCORE* IP Block Memory Generator. The RAM core has a word width of 16 bit and a word depth of 125,952, which is the maximum number of cells for a single RAM core on the FPGA. It is a simple dual port RAM, which means it has dedicated write and read ports with their own clock and address ports. Both ports are clocked with the internal 27 MHz clock.

1.6 Reading and Sending Audio Data

Any audio data that is to be transmitted from the Armada extension board to the FPGA has to be formatted in a 16-bit PCM WAV format. PCM WAV specifications can be found at [2]. The program first reads the header of the file, and then all samples in 16-bit chunks. Some files may have additional header-like data at the end, which is read as normal audio data, so it has to be stripped beforehand using e.g. a hex editor. After having stored the audio data in an array, the program then uses I^2C library functions to transmit the data to the FPGA where it is stored in the block RAM and played. To indicate the end of data, the program sends a 0xAAAA (1010101010101010) after the audio data.

1.7 Future Work

On the FPGA side of the project, further work should include using more than one of the chips block RAMs to be able to intermediately store more data at a time. This should be an extension of the already present code, theoretically using an array of multiple instances of the already generated IP core should suffice. The goal should be to use the DDR2 RAM to buffer data from the SD card and directly play from that.

Additionally, some more controls for the playback can be implemented, in addition to the already available volume control. Pause/Play and even fast forward/reverse are possible functions. Concerning the C code that is executed on the Armada board, it would also be an advantage being able to load other formats than PCM WAV.

2 VHDL Based Spectrum Analyzer

The goal of the presented project is to display spectral data of an audio-input via a HDMI-output and to change different parameters like the window function or the color with a universal remote control. All needed modules are implemented in VHDL and they are running on the Xilinx Spartan 6 FPGA on E2LP board. The HDMI and remote control part is described in Chap. 3.

An overview of the hardware connection is shown in Fig. 1. To get sampled audio data from an analog source the ADAU1772 audio chip from Analog Devices, which is also located on the E2LP board, is used. So the first part of the project was to configure this chip properly. Therefore an I^2C module has been created. After that the sampled audio data, which are provided from the audio chip over an I^2S stream, must be read in and outputted again. To do that an I^2S module was written in a second step. The third and biggest step was to build a Fast Fourier Transform to get the spectral data out of the sampled audio data. In the following these three steps/modules are described in more detail.

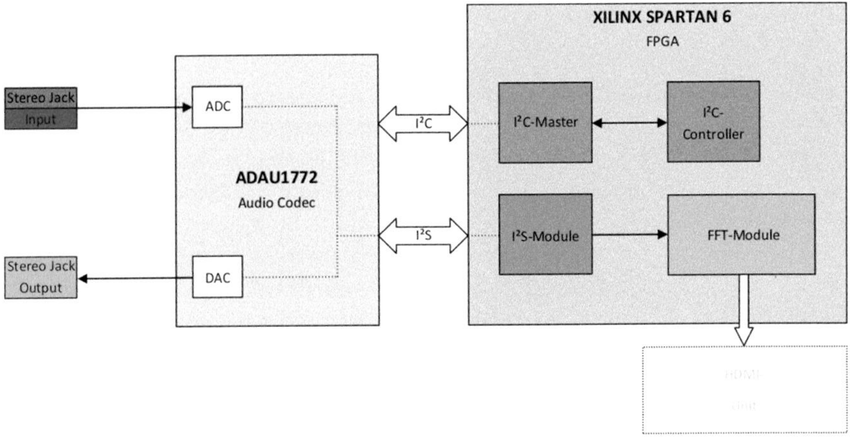

Fig. 1 Connection diagram of hardware used on E2LP board for audio analysis

2.1 Configuration of I²C Module

Figure 2 shows the I²C module in more detail. It consists of two parts, the first one is an I²C master which was made up especially for the needs of the ADAU1772 chip [3]. This module provides the direct communication with the chip via the SDA and the SCL lines and is capable of reading and setting all registers, parameters and storages on the chip. The second module is the I²C controller which sets about 24 registers (all extracted from [3]) to setup the chip for a proper use. Therefor it transfers to the master module a register-address, a read or write command and if necessary the data which should be written and then starts the execution with the Start signal. Additionally the audio chip is powered on in this module and two bits of the I²C-chip-address are set here.

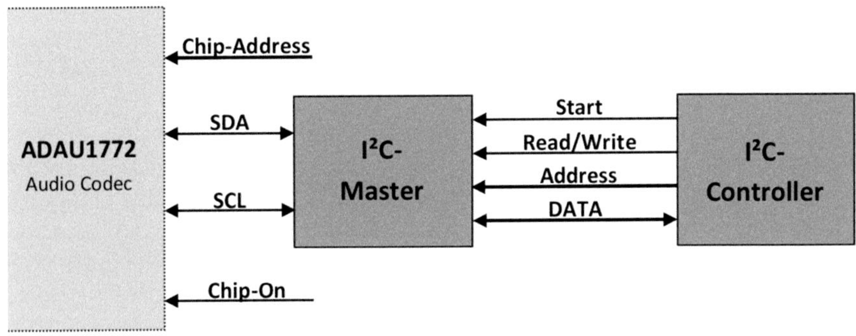

Fig. 2 Connection diagram of I²C module

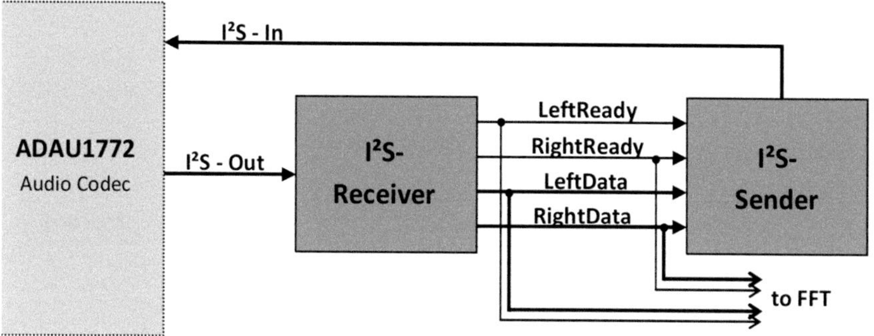

Fig. 3 Connection diagram of I^{2}S module

2.2 *Configuration of I^{2}S Module*

The I^{2}S module on the board consists of two parts as shown in Fig. 3. The first one is an I^{2}S receiver. It receives an audio (stereo) I^{2}S stream from the audio chip and extracts the 24-bit sampled data from the left and the right audio channel separately. Each time the module has received a complete sample the data is outputted via the data line and the corresponding ready line is triggered. These outgoing lines are routed to the Fast Fourier Transform module on the one hand and to an I^{2}S sender on the other hand. The I^{2}S sender directly returns the audio data samples to the audio chip where they are converted back to analog signals and outputted again. With this structure it is possible to loop a stereo audio signal through the E2LP board and so its spectral data could be shown while listening, without a special wiring effort.

For both modules, the receiver and the sender, the I^{2}S clock is provided by the audio chip which is therefor the master. Since the sampling rate is set to 48 kHz approximately every 21 μs a new sample is ready.

2.3 *Fast Fourier Transform Module*

The key part of the audio-analyzer is the Fast Fourier Transform (FFT) module, which is shown in Fig. 4. The implementation of this module was inspired by [4]. For each audio channel (left and right) there is a separate FFT module. The FFT-Handler is a big state-machine which is triggered by the ready signal from the I^{2}S module. If the data is ready the upper 11 bit of a sample are stored for further processing. Since the implemented FFT is a 16-point-FFT this is done 16 times. After that the stored data is processed according to the butterfly structure. Therefore always two of the data values and a complex twiddle factor are fed through the butterfly unit. The two results are stored and again processed in the

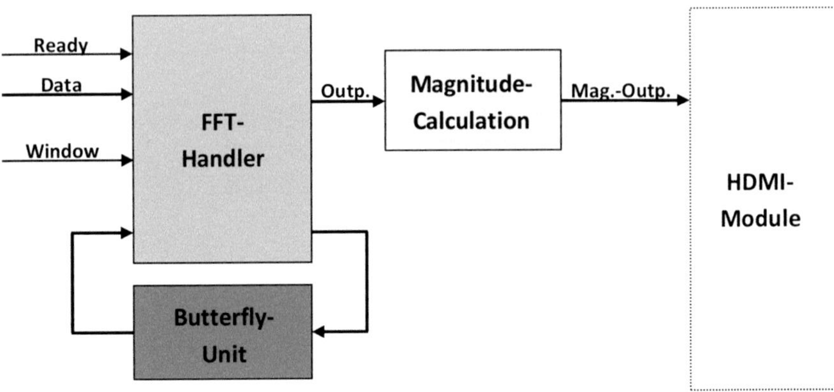

Fig. 4 Diagram of the FFT-module including a Butterfly unit

same way until the four FFT levels are completed. The final, now complex, values 1–7 (these are the only meaningful) are then outputted into the magnitude-calculation module where the magnitude is calculated. The magnitudes are then displayed via the HDMI module.

Since the Butterfly-Unit has a latency of 5 clock-cycles a nearly maximum pipelining is implemented. To do that and to avoid data hazards at the beginning of every FFT level one data package per clock cycle is shifted into the unit and the next level is not started until all results are received. Also the magnitude calculation module has a latency of 11 clock cycles. To avoid long waiting times the magnitude calculation is therefore done in parallel to the storage of the samples, because this is relatively slow by nature (48 kHz).

To receive a meaningful result from the FFT a windowing of the input-samples can not be omitted. So as a additional part the capability of selecting different windows was implemented. This is done by a multiplication of the samples with a weighting factor gained from a window function before storing. Via the Window signal it is possible to select between No Window, Von-Hann Window, Hamming Window and Blackman-Harris Window. The currently selected window is displayed on the display of the E2LP board by an extra module and it could be selected with a remote control. This additional feature allows one to easily explore the impact of the different window functions.

The presented FFT module delivers good results and with the pipelining it is able to perform in real-time (which means a whole FFT calculation can be done between two samples when a sampling rate of 48 kHz is used). The module also provides the ground structure for FFTs of a higher order (e.g. 32-, 64-, or 1024-point) which could simply be created by adding more Store- and Butterfly-states to the state machine.

3 FFT—Spectral Data Output via HDMI Port

The aim of this project is to display the spectral data from a FFT on a HDMI-Display and to control different functionalities with a universal remote control. In the following all used VHDL entities are briefly described. The connection diagram of the modules is graphically displayed in Fig. 5.

3.1 Module Configuration Over I^2C

The HDMI Chip ADV7511 requires at least 16 registers to be set [5]. Most of the data that has to be set consists of manipulating single bits in the 8-Bit registers. For this task an I^2C-Master that operates on 100 kHz clock has been implemented. For the control of this Module a controller had to be set up. Using these two modules it is possible to read data from the chip, manipulate single bits and write data back to the IC.

The I^2C-Master waits in its Idle-state for the commands, holding the READY-Output on '1'. For writing a slave address on the bus a controller has to set the slave's address to the SLAVEADDRESS-Input, a '0' for writing or '1' for reading to the ROW-Input, an "01" to the ARW-Input (Address, Read, Write) and set the SENDDATA-Input to '1' to start the transmission. In the next clock-cycle the I^2C-Master will set the READY-Output to '0' to signal that it's busy. The SENDDATA-Input must be held on '1' during the whole command, otherwise a STOP-Condition will be created with misreporting the slave that a transmission is already finished.

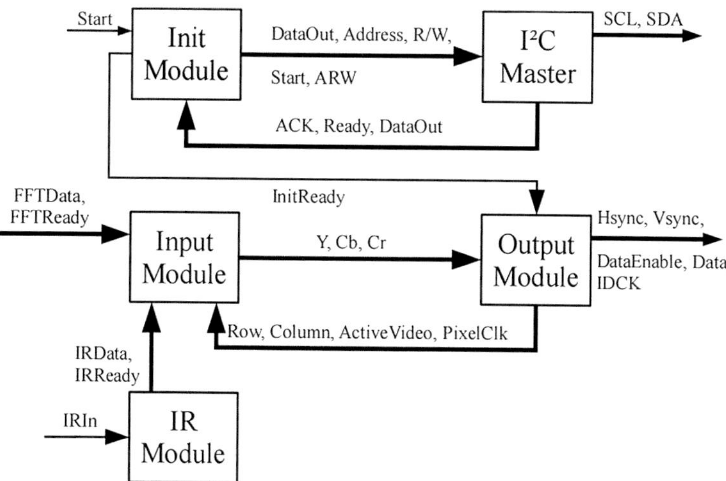

Fig. 5 Module connection diagram for data display

After the slave's address has been sent, data can be written to the slave. If data is to be written, the controller must set the data to the DATAIN-Input and set the ARW-Input to "11". The transmission will take place, after the I C-Master has signaled a ready and the SENDDATA was on '1'.

For a read sequence the controller must, after sending the address, set the ARW-Input to "01" and keep the SENDDATA-Output on '1'. After the data has been read from the slave it appears at the DATAOUT-Output of the I C-Master. In all cases the slave's acknowledge appears at the ACKOUT-Output, so that the controller can react accordingly. If the slave did not acknowledge a transmission (acknowledge bit = '1') the I^2C-Master sends (according to the I^2C-Standard) a STOP-Condition.

3.2 HDMI Initialization and Control Over I^2C

For setting up the ADV 7511 W an initialization module is to be set up, that uses the implemented I^2C-Master to set several registers on the IC. In [5] the registers that must be set for proper operation are described (0x41 – 0xF9). Besides these registers, others are set as described in Table 2. The Initialization-Sequence is started using the lower button on the front-panel of the E2LP-Board (Table 2).

3.3 Displaying the FFT-Spectrum with HDMI

For displaying the incoming data of the FFT two modules have been set up. One module takes the input from the FFT and creates a bar pattern, in which the height of the bars corresponds to the input value of the channels. In this module also the color of the bar pattern can be changed using the remote control. The other module handles the controlling of the ADV 7511 W based on the input from the input module.

Table 2 Register configuration of ADV 7511

Register address	Value	Effect
0x15(3:0)	0001	Input-Id = 1
0x16(5:4)	11	Color-depth = 8 Bit
0x16(7)	0	Output-format 4:4:4
0x16(3:2)	01	Input-style 2
0x16(0)	0	Output-color-space (for black image) = YCbCr
0x17(1)	0	Aspect-ratio: 4:3
0x18(7)	1	CSC enabled
0xAF(1)	1	HDMI-Mode
0x97(1)	1	DDC controller interrupt (probably not necessary)

In the input module the video input is created. In the module the current pixel position from the ROW- and COLUMN-input is compared with the position of the bars and either black or the color of the corresponding bar (color 1 to color 14) is set to the color output. Each bar has a start- and an end-value, that determines in which column the bar starts and in which it ends. The height of the bars is determined by the 17 Bit wide BAR1 to BAR14 inputs. Their input is shifted 7 Bits to the right and thus linearly mapped (division by 128) to a 8 Bit value. Furthermore in this module the input from the IR-module is used to set the color of the bars. With button 4 one can switch between red, green, yellow, white and mixed colors for the left block of bars. With the button 5 one can do this for the right block. With the button 6 and 7 one can add the value 5 to each color value (Y, Cb, Cr) to obtain other colors for the left and right block.

In the output module the handling of the ADV 7511 W is done by creating The HSYNC, VSYNC, DATAENABLE, DATA and IDCK-Signals according to the ITU BT.1358-standard described in [6, 7]. The used video mode is 480p. By using other constants for the timing and supplying a higher clock rate one could also use higher resolutions. It is also important to know that the clock-rate to the ADV 7511 W is twice the pixel clock which is created in this module [5]. The double data clock rate is created with a *Xilinx Digital Clock Manager* IP-Core.

3.4 IR Remote Reading—IR Decoder

Another feature of the realized project is to read and decode the signal from a Universal remote control to control different functionalities. The data that is sent is encoded in RC5-format using Manchester-Coding as described in [8]. One data frame consists of 14 Bits which are sampled using a state machine that uses the length of the high- and low-phases and the edges to decode the data. If a complete data-frame was sampled the OUT-Output together with a '1' at the DATAREADY-data appears on the DATA-output. To use the RC-5-coding the remote control is programmed to a SEG-TV.

4 IMU—Fancy Box Project

4.1 Hardware and Software Design

In this project the whole system is divided into smaller tasks implemented in VHDL running on an FPGA and tasks implemented in the high level language Java running on a typical CPU. The IMU part is done in VHDL to make use of the parallel execution and data processing. Data of the rotation speed is gathered at maximum speed and is processed in real time to minimize drift and offset errors. With a typical CPU running a non-real-time operating system, time critical data

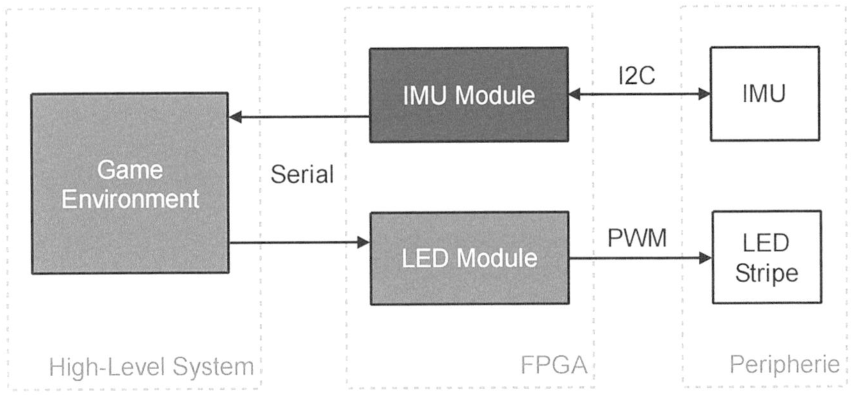

Fig. 6 Schematic block diagram of data processing in eddy current tomography

processing cannot be done without significant errors. The gathered, processed and filtered data is afterwards transmitted to the high level operating system via a serial connection. A connection overview is shown in Fig. 6. The game physics and screen output are implemented in software because these tasks are not time critical and are much faster and easier implemented in high level languages like Java. Since the transmission of the color data to the LEDs is done with *pulse width modulation* which is again time critical, it is done on the FPGA as well.

4.2 IMU Data Acquisition with I^2C

The IMU is connected over an I2C bus to an I2C master module like shown in Fig. 7. This is written modularly and can be used for other tasks or projects as well. An IMU controller sets the register addresses and data accordingly, it sets the

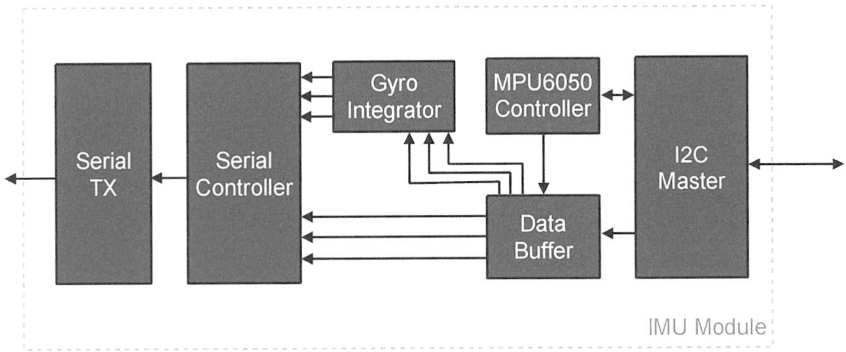

Fig. 7 Schematic of the IMU VHDL modules

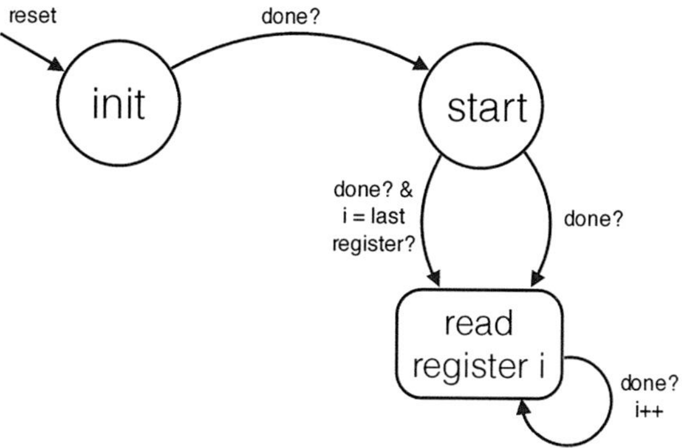

Fig. 8 The state machine of the IMU controller module

IMU active and repeatedly asks for the angle values at the maximum IMU speed of 400 kHz. This is implemented with a simple state machine like shown in Fig. 8.

In the init state a '0' is send to the power management register of the IMU. This forces the IMU to start angular measurements. In the start state the first register address is sent to the IMU together with a continuous read command. The registers are afterwards read out in a loop at maximum I^2C speed. After a complete set of angle data is read, the gyroscope data is filtered and integrated. The first 1024 data samples are used to calculate an offset which is further subtracted from every sample. The result is integrated following Eq. 1 afterwards.

$$\text{gyro\_integrated} \Leftarrow \textit{to\_unsigned}(\text{gyro\_integrated}) + \textit{to\_unsigned}(\text{new\_gyro})\;; \quad (1)$$

Since the integrated gyro value is limited (in this case to 32 Bit) the maximum angle of turn is fixed (at around 6600° which correspond to approx. 18 turns in a specific direction). After the data is processed, it is send over a serial connection to a high level system where the data of the gyroscope and accelerometer is converted into system angles with the following Eq. 2.

$$\text{angle} = a \cdot \text{gyro\_integrated} \cdot dt \cdot \frac{1}{\text{gyro\_sensitivity}} + (1-a) \cdot \text{acc\_angle} \quad (2)$$

The influence of the accelerometer on the angle calculation can be weighted with the complementary factor a.

4.3 Java Game with a Moving Ball

By tilting the IMU a virtual ball can be controlled in a Java game application by applying virtual gravity to it. The look and design of the game is shown in Fig. 9.

The goal of the game is to destroy all boxes. Furthermore it is only allowed to destroy the box with the same color as the ball. After the box is destroyed, the ball switches to a color of the next box that needs to be destroyed. The number of boxes is variable and their positions and colors are random.

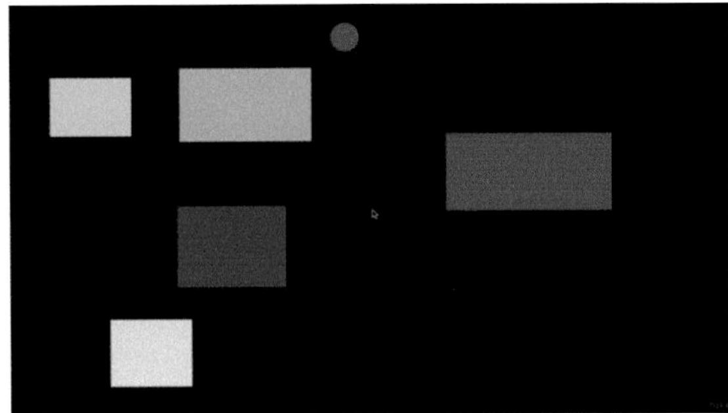

Fig. 9 The look of the final Java based game

4.4 Add Color to Your Life—VHDL Based LED Controller

The data is buffered in the memory and the LED controller is notified that new data is available. The LED controller sends the RGB data of all LEDs to the WS2811 controller module which converts the Data to the corresponding PWM signal.

References

1. Analog Devices, Four ADC, Two DAC Low Power Codec with Audio Processor. Data Sheet, http://www.analog.com/static/imported-files/data_sheets/ADAU1772.pdf
2. WAVE File Format Specifications. McGill University, Montreal, Canada. http://www-mmsp.ece.mcgill.ca/Documents/AudioFormats/WAVE/WAVE.html (2014)
3. Analog Devices. Datasheet—ADAU1772 Audio Codec (2014)
4. William Slade, G.: The fast fourier transform in hardware: a tutorial based on an FPGA implementation (2013)
5. ADV7511W Low-Power HDMI 1.4 A Compatible Transmitter PROGRAMMING GUIDE. http://www.analog.com (2012)
6. Jack, K.: Video demystified—a handbook for the digital engineer. Elsevier, Burlington
7. CEA Standard, A DTV Profile for Uncompressed High Speed Digital Interfaces. Technology and Standards Department, 2500 Wilson Boulevard, Arlington, VA 22201 (2006)
8. Data Formats for IR Remote Control. http://www.vishay.com

E2LP Remote Laboratory: Introduction Course and Evaluation at Warsaw University of Technology

Rafał Kłoda, Jan Piwiński and Roman Szewczyk

Abstract Paper presents the results of Remote Laboratory services application into Embedded Systems Education, developed under Embedded Computer Engineering Learning Platform FP7 project. This paper reports on results of performed Remote Laboratory evaluation in Warsaw University of Technology, where we introduce the new learning model in Digital System Design course. Students got an introduction to Xilinx software environment, exercises laboratory example of fundamentals in VHDL programming language and introduction to different digital logic circuits and their operation. The paper presents the analysis and results of performed evaluation based on Computer System Usability Questionnaire. For data analysis R programming language and software environment for statistical computing and graphics was used.

Keywords E2LP · Curriculum · Evaluation · Remote laboratory

R. Kłoda (✉) · J. Piwiński
Industrial Research Institute for Automation and Measurements,
Al. Jerozolimskie 202, 02-486 Warsaw, Poland
e-mail: rkloda@piap.pl

J. Piwiński
e-mail: jpiwinski@piap.pl

R. Szewczyk
Institute of Metrology and Biomedical Engineering,
Warsaw University of Technology, ul. sw. A. Boboli 8, 02-525 Warsaw, Poland
e-mail: szewczyk@mchtr.pw.edu.pl

R. Szewczyk et al. (eds.), *Embedded Engineering Education*,
Advances in Intelligent Systems and Computing 421,
DOI 10.1007/978-3-319-27540-6_9

1 Introduction

Embedded Computer Engineering Learning Platform (E2LP) is a European FP7 project of 3 years duration, started in September 2012 [1]. The project deals with Embedded Systems Education at University level, which requires a multidisciplinary approach, involving different technologies and system solution optimizations.

The main purpose of the project is to introduce the new learning model and innovative teaching methods in Embedded Engineering, which cover following challenges:

- Assembling advanced hardware for E2LP platform that consists of a low cost Spartan-6 Platform FPGA with comprehensive collection of peripheral components that creates a complex and unified embedded system.
- The set of exercises laboratory examples to teach students the fundamental concepts in FPGA-based embedded system design, which consider skills practice through supporting individualization in learning.
- Integration of Augmented Reality (AR) interface for visualizing, simulating and monitoring invisible principles and phenomena in the field of embedded electronics.
- Provide Remote Laboratory (RL), which provides an access to continuously refreshed interfaces and signals from the real board and enable user remotely control and program the board directly from their computer at home.
- Evaluation of Learning Technology, which deals with cognitive theories on how people learn and will help students to achieve a stronger and smarter adaptation of the subject.

To reach the aforementioned challenges we wanted to test the developed Remote Laboratory before the official system evaluation in curriculum of E2LP project academic partners. The approach was to introduce the selected set of exercises from Digital System Design course as a supplement to Intelligent Measurement Devices course, lead by Prof. Roman Szewczyk in Warsaw University of Technology. Hereunder we present the methodology and results of evaluation as well as student's feedback toward system.

2 Evaluation and Discussion on Results

This paper presents the preliminary student's practical validation of developed E2LP Remote Laboratory [2], which was performed at Warsaw University of Technology (WUT) at Mechatronics Faculty during the evaluation stage of the E2LP project. The main purpose of performed evaluation was showing to students system capabilities and engaging them in contribution in testing the developed RL platform as a additional value to study programs in WUT.

A study was completed under "Intelligent Measurement Devices"—a new course in the Electronic Measurement Systems specialization on engineering degree.

During this course students gain comprehensive skills: knowledge about the intelligent sensors, measurements devices and systems operation rules, competence in signal processing and the methodology of novel apparatus construction.

One of the main purpose of the evaluation and the E2LP project was the enrich aforementioned skills with understanding the different digital logic circuits and their operation, implementation of Boolean functions using digital logic circuits, understand the Xilinx ISE software environment and tools as well as understand VHDL description of digital logic circuits.

To get the summative feedback from students towards presented system, the quantitative on-line analysis (Table 1) was conducted, which was prepared by other E2PL project partner Ben-Gurion University of the Negev, based on Computer System Usability Questionnaire (CSUQ) [3]. It includes many aspects that refer to the usage RL platform and its and user acceptance.

Our E2LP Remote Laboratory is innovative learning platform, which easy customizes to any course needs and doesn't require any cost for teachers, namely they don't need any specialize software and hardware. Since the students were

Table 1 Evaluation questionnaire for remote laboratory with summarized results

Question	Description	Mean	Std. Dev.
1	Overall, I am satisfied with how easy it is to use this system	2.9	1.50
2	It is simple to use this system	3.0	1.55
3	The interface of this system is pleasant	3.1	2.00
4	I am able to complete my work quickly using this system	3.2	1.73
5	It is easy to find the information I need	3.6	1.75
6	I am able to efficiently complete my work using this system	3.5	1.51
7	I feel comfortable using this system	3.4	1.50
8	The information (such as on-line help, on-screen messages and other documentation) provided with this system is clear	3.6	1.78
9	I Believe I became productive quickly using this system	3.4	1.41
10	The system gives error messages that clearly tell me how to fix problems	4.5	1.51
11	I can effectively complete my work using this system	3.6	1.55
12	Whenever I make a mistake using the system, I recover easily and quickly	3.6	1.36
13	The information provided with the system is easy to understand	3.6	1.59
14	I like using the interface of this system	3.5	1.59
15	The system information (such as on-line help, on-screen messages and other documentation) is effective in helping me complete my work	3.5	1.76
16	The organization of information on the system screens is clear	3.0	1.63
17	This system has all the functions and capabilities I expect it to have	3.6	1.71
18	It was easy to learn to use this system	3.3	1.66
19	Overall, I am satisfied with this system	3.2	1.44

beginners in VHDL language and Xilinx environments we prepared separate set of easy exercises, which were in line with the curriculum of the subject.

Using RL system students could remotely program given set of exercises over the Internet and simultaneously, in real time could monitor the evolution of the experiment on dedicated Graphical User interface (GUI) of the Front Panel of the real board [4]. Students after successful finalization of all exercises were asked to fill prepared in system special questionnaire. Their were required to rate the prepared questions on a 7-point Likert scale (7—strongly disagree and 1—strongly agree) [5].

3 Questionnaire Data Analysis

The statistical study was conducted in the fifth semester of the academic year 2014–2015 on one group of users (16 students of this course, who are not specialists in this field). The analysis of the results from this study was performed using statistical environmental named "R" [6]. The histograms of collected data are presented in Fig. 1.

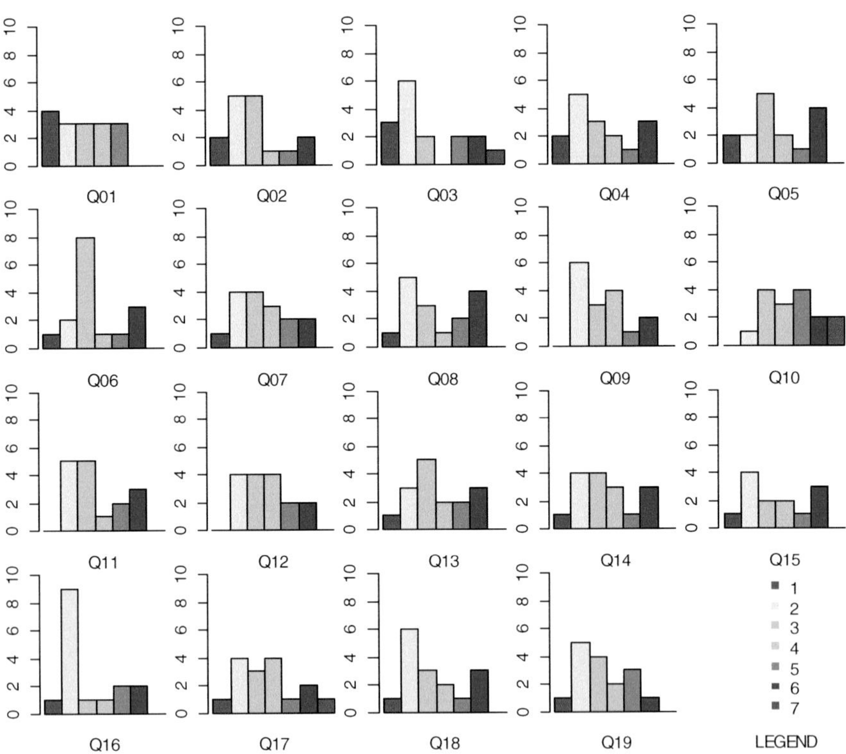

Fig. 1 Distributions of collected rating

Since students knowledge about FPGA systems was very low there are many disparities between their individual responses for given questions (Fig. 1). Therefore of strong differences between individual response on the same question, standard deviation for average scores is large (Table 1). This to some degree, is caused by the different users needs from the system: their requirements and expectations (Fig. 1, Q17), but also their ability to learn (Fig. 1, Q18), and overall the interpretation of the semantic scale used in evaluation survey. At this point it should be noted that the end point on Likert scale (7–strongly disagree) practically are not used for the rating. This could be caused by lack of student's references to other systems. For these particular students the E2LP Remote Laboratory was the first contact with this specific software as well as FPGA platform.

However obtained results are very valuable observations. The average responses obtained in our analysis are shown in Fig. 2.

Regarding the e-learning platform for FPGA, the users confirm, that proposed solution are powerful and efficiently improved by using RL (questions 1, 2, 7 and 19). In this sense, students declare they somewhat agree with the idea that remote work is possible without the need to work with the real board (questions 1 and 6).

Considering the aspects related to the user graphic interface, users proof that it is easy to use and its readability increase significantly (questions 3, 5, 14, 16 and 18). Moreover students stress that they are able to quick learning (questions 4, 9, 11 and 18).

The biggest encountered problem was connected with using Xilinx ISE software, which was source of error messages (question 10) and poor programming experience (questions 8, 12, 13 and 15). This mark might be a reason of very low student's initial knowledge level of FPGA systems.

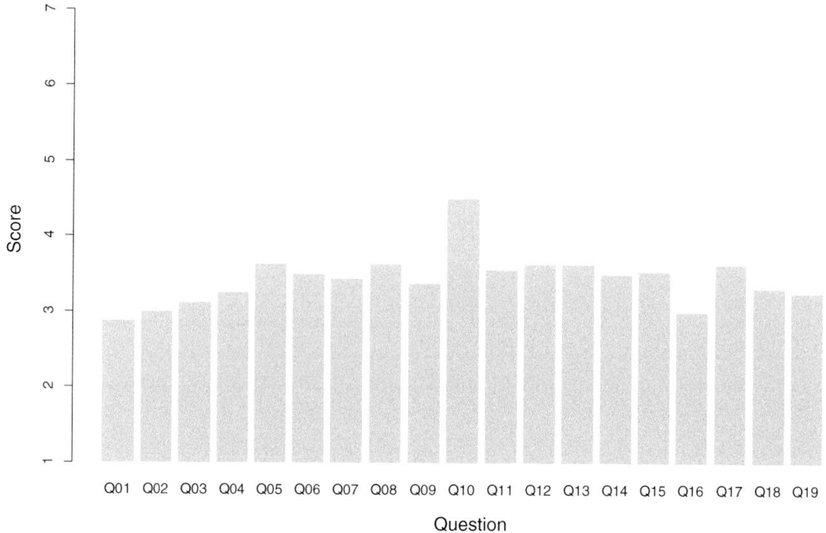

Fig. 2 Averages responses from students

4 Conclusions

Results presented in the paper confirms that introduction of RL into curriculum and new learning model is challenging in the education of engineers in embedded systems. Student, who has never had any practice with Xilinx ISE environment and any FPGA board configuration needs really precise procedure what to do in current exercise.

Remote laboratory enable users to access E2LP platform over the Internet, configure it compiling VHDL code and having the immediate feedback of solution on their own computer.

During the evaluation it occurred that remote operations through real-time experiments stimulate the students curiosity and productivity, which was presented in aforementioned questionnaire results. It also occurred to us that ascending order for marks where 7 is strongly disagree, opposite to Polish degrees order, might have caused misunderstanding, or simply overlook due to habits, so consequently disturb the questionnaire results.

Acknowledgments E2LP Remote Laboratory evaluation were made in the Institute of Metrology and Biomedical Engineering, Warsaw University of Technology.

References

1. E2LP Project Website. http://www.e2lp.org
2. E2LP Remote Laboratory, e-learning portal for E2LP. http://e2lp.piap.pl
3. Lewis, J.R.: IBM computer usability satisfaction questionnaires: psychometric evaluation and instructions for use. Int. J. Hum. Comput. Interact. **7**(1), 57–78 (1995)
4. Kastelan, I., Benito, J.R.L., Artetxe Gonzalez, E., Piwinski, J., Barak, M., Temerinac, M.: E2LP: a unified embedded engineering learning platform. Microprocess. Microsyst. Elsevier, **38**(8), Part B, 933–946 (2014)
5. Wade, V.M.: Likert-type scale response anchors. Clemson International Institute for Tourism & Research Development, Department of Parks, Recreation and Tourism Management, Clemson University (2006)
6. R Development Core Team: R: a language and environment for statistical computing, R Foundation for Statistical Computing, Vienna, Austria. ISBN 3-900051-07-0. http://www.R-project.org (2010)

Exploring Aspects of Self-regulated Learning Among Engineering Students Learning Digital System Design in the FPGA Environment—Methodology and Findings

Moshe Barak, Ivan Kastelan and Zvi Azia

Abstract This study addressed the case of the development, implementation and evaluation of an innovative platform for teaching computer and embedded systems engineering, including hardware, software and instructional materials for students. The evaluation methodology was derived from the Self-Regulated Learning Theory which relates to cognitive, meta-cognitive, motivational and self-efficacy aspects of learning. Data were collected by administrating the Lab Feedback Questionnaire, the Motivated Strategies for Learning Questionnaire (MSLQ), the Computer System Usability Questionnaire (CSUQ), and interviews held with students and teachers. The findings that were obtained over two years taught us that one of the most important factors affecting the success of an advanced technological learning platform is the careful design of students' assignments, for example, to progress gradually from solving basic exercises, to solving more significant problems and dealing with broad projects.

Keywords Computer and embedded engineering · Self-Regulated Learning · Exercises, problens and projects

M. Barak (✉) · Z. Azia
Department of Science and Technology Education, Faculty of the Humanities and Social Sciences, Ben-Gurion University of the Negev, Beer-Sheva, Israel
e-mail: mbarak@bgu.ac.il

Z. Azia
e-mail: zvi.azia@gmail.com

I. Kastelan
Department of Computing and Control Engineering, Faculty of Technical Sciences, University of Novi Sad, Novi Sad, Serbia
e-mail: ivan.kastelan@rt-rk.uns.ac.rs

R. Szewczyk et al. (eds.), *Embedded Engineering Education*,
Advances in Intelligent Systems and Computing 421,
DOI 10.1007/978-3-319-27540-6_10

1 Introduction

Digital system design is one of the basic subjects learned in a number of engineering areas, for example, electrical engineering and computer engineering. The traditional method of teaching this subject is conducting 'chalk and talk' lectures and involving students in standard pre-designed experiments based on modular lab equipment. This method is very outdated in light of the rapid development of digital technology and the industry extensive transition to using reprogrammable devices. Using simulation instead of involving students in hands-on work prevents learners from gaining real engineering experience.

This paper addressed the case of the development, implementation and evaluation of the Embedded Engineering Learning Platform (E2LP), an innovative low-cost learning environment for computer and embedded engineering that was developed by a consortium of nine European partners [1]. The E2LP Base Board shown in Fig. 1 is based on a low-cost FPGA, Xilinx Spartan-6, surrounded by a comprehensive collection of peripheral components that can be used to create a complex system, and a learning platform that covers most laboratory tasks in computer and embedded engineering courses. The board enables students to learn about FPGA, digital design and computer architecture.

The system also includes two extension boards. One is the ARMADA, which is based on the dual-core ARM A9 SoC, a 32-bit processor core. Among the other components installed on the ARMADA are a state-of-the-art DSP and peripherals. The ARMADA board allows executing more complex experiments involving advanced computer architecture, multimedia and OS implementations. The second extension board, NXP, is based on the ARM7 processor, which is a simple CPU. Typical sensors, such as a thermometer and an accelerometer, are installed on the board for basic programming exercises such as typical control applications. The main board with the two extension boards can be used for teaching subjects such as digital systems, control systems, communication systems and digital signal analysis to students in a range of engineering fields, for example, computer and embedded engineering, electrical engineering or mechanical engineering [2].

In addition to the hardware and software development, the E2LP project also included the preparation of instructional materials for students and teachers, as

Fig. 1 E2LP Base Board

well as experimental applications of the platform in a wide range of subjects. In the years 2014–2015, the platform was used for teaching courses such as Logic Design of Computer Systems, Digital Signal Processing, Computer Networks, Real-Time System Software, Advanced Embedded Systems, Computer Architecture and Mechatronics Control. These courses took place in universities in Serbia, Croatia, Germany, Poland and Israel, and involved 630 students and 50 teachers.

The project adopted the internal formative evaluation approach, which, according to Owens [3]:

- Involves the staff as much as possible;
- Strives for consensus on the evaluation plan;
- Encourages evaluators to report on their progress;
- Uses findings to reflect on the program aspects under review; and
- Develops a systematic plan by which changes can be made.

This paper aims at examining the implementation of the platform in a digital system design course in one of the universities, with focus on students' achievements and motivation, through the lens of a comprehensive evaluation methodology that was developed within the project.

The study was guided by two main questions:

1. Which components of the new learning platform and the instructional materials supported or hindered students' success and motivation in learning the courses?
2. To what extent and how did the evaluation methodology applied in the project provide effective feedback to the teachers and the developers and contribute to the project's success?

2 Literature Review

Education in general, and engineering education in particular, is not only about imparting specific knowledge to students, but also about developing students' thinking and learning skills, as well as their motivation to learn. To achieve this end, the E2LP project adopted a broad conceptual framework for curriculum development and evaluation of teaching and learning. This section highlights two major concepts that guided the evaluation methodology: "Self-Regulated Learning" and "Task Taxonomy."

2.1 *Self-regulated Learning*

The term Self-Regulated Learning (SRL) is defined as the active, goal-directed, self-control of behavior, motivation and cognition for academic tasks [4–6]. Below we refer to the three main dimensions of SRL: cognition, meta-cognition and motivation.

- The term "cognition" relates to the conscious mental processes by which knowledge is accumulated and constructed, understanding, problem solving and creative thinking.
- Metacognition refers to 'thinking about thinking,' knowledge of general strategies that might be used for different tasks, and the process of selecting, controlling or regulating cognitive processes such as learning and problem solving before, during or immediately after executing a task [7, 8]. Metacognition dimension also includes reflective practice—the process of learning from experience, asking questions about what we know and how we came to know it, and 'learning to learn' [9].
- The motivational dimension in the SRL model includes aspects such as interest in learning, intrinsic motivation, extrinsic motivation and self-efficacy beliefs [10, 11].

Pintrich [8, p. 401] writes: "The self-regulated learning does provide a conceptual model of college student motivation and regulation that is based in a psychological analysis of academic learning. In addition, there is fairly wide empirical support from both laboratory and field-based studies for SRL models of this type." In recent years, the SRL model caught the attention of a growing number of researchers in engineering and science education [12–14].

2.2 The Task Taxonomy

In the educational literature, there is a wide consensus that one of the major factors affecting students' learning is the type of tasks the learners deal with in class. This view influenced the increasing emphasis of introducing more learner-centered instructional strategies, such as authentic learning, problem-based learning and project-based learning into science and technology education [15, 16]. However, as Kirschner et al. [17] show, learners need some preparation and basic guidance before being involved in project-based learning. From this perspective, we developed the Task Taxonomy, according to which it is useful to distinguish between three levels of students' assignments:

- Exercises: closed-ended tasks in which the solution is known in advance and the learners can check if they have arrived at the correct answer.
- Problems: open-ended, small-scale tasks in which students might use different solutions and methods or arrive at different answers.
- Projects: challenging tasks in which the problem is ill-defined. Students take part in defining the problem, setting objectives, identifying constraints, and choosing the solution method.

The SRL theory and Task Taxonomy presented above provide a broad view of students' achievements, learning competences and motivation to learn a new subject. These concepts were also used for the evaluation of students' learning in computer engineering courses [1, 18] and in exploring technology education in high schools [16]. In the current study, these concepts guided the course design and evaluation, as presented in the following sections.

3 Setting

3.1 Participants

The study took place among students and teachers at the Faculty of Technical Sciences, University of Novi Sad, Serbia. About 220 students and nine teachers were involved in each of the two cycles of learning the Digital System Design 1 course during the years 2014–2015.

3.2 The Course Program

The Digital System Design 1 course is a basic course in computer and embedded systems engineering that combines lectures and lab work based on the E2LP main platform. The students carried out the following lab experiments, which were developed by the consortium:

- Lab 1—Digital Logic Circuits
- Lab 2—VHDL Gate-Level Design of Digital Circuits
- Lab 3—Combinational and Sequential Circuits
- Lab 4—Problem Set: Combinational Circuits
- Lab 5—Problem Set: Sequential Circuits
- Lab 6—Finite State Machines and Complex Digital Systems
- Lab 7—Problem Set: Finite State Machines
- Lab 8—Problem Set: Complex Digital Systems
- Lab 9—Computation Structures
- Lab 10—Project: CPU Design, part 1—Computation Circuits
- Lab 11—Project: CPU Design, part 2—Control Unit
- Lab 12—Project: CPU Design, part 3—Memory and Programming

According to the Task Taxonomy mentioned above, labs 1–3 were defined as basic exercises, labs 5–9 as problems, and labs 10–12 as projects. The evaluation program aimed at exploring students' success and motivation in learning the course and performing the lab tasks, as described in the following sections.

4 Method

4.1 Research Methodology

The current study adopted a mixed approach and combined quantitative and qualitative methods. Authors such as Olds et al. [19], and Borrego et al. [20] address the applications of using mixed research methods in the context of engineering education. The quantitative side often consists of having the students' answer questionnaires or exams. Qualitative data, often obtained by observations and interviews, can provide information about the quality of standardized case records and quantitative survey measures, as well as offer some insight into the meaning of particular fixed responses. Koro-Ljungberg and Douglas [21] also point out that increased use of qualitative methods in engineering education research may allow new understandings to emerge and contribute to the study of complex problems and socio-cultural phenomena that cannot be answered through qualitative methods.

4.2 The Lab Feedback Questionnaire (LFQ)

In order to examine students' success and motivation in carrying out the lab experiments, we developed the short Lab Feedback Questionnaire (LFQ), which the students filled in about five times during the semester. The students marked their answers on a Likert scale (1-very low… 10-very high) to the following eight questions:

Clarity of theoretical background—documentation, theoretical explanations.
Clarity of technical instructions, exercises and problems.
Total time and efforts required.
Ease of use of the environment—software issues.
Ease of use of the platform.
Assistant support—was the lab assistant helpful.
To what extent do you think you learned something valuable?
Overall satisfaction.

The students were also asked to explain or give examples of their answers in an open text box, as illustrated in the example shown in Fig. 2 and Appendix 1.

4.3 The Motivated Strategies for Learning Questionnaire (MSLQ)

At the end of the semester, the students answered the Motivated Strategies for Learning Questionnaire, which had to do with measuring aspects of motivation and metacognition according to the Self-Regulated Learning (SRL) theory guiding the evaluation in E2LP. The full version of this questionnaire, which is well-known

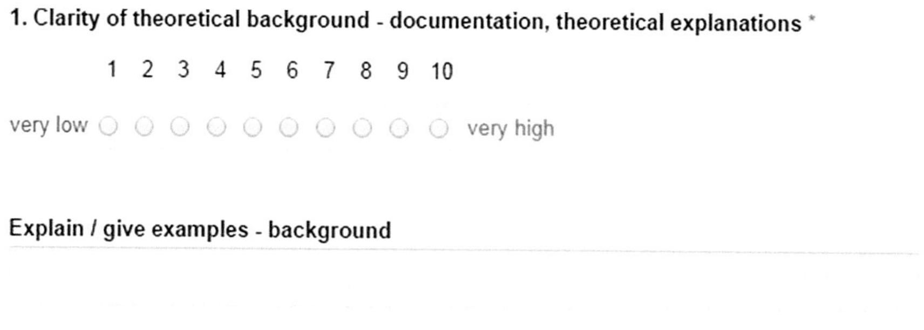

1. Clarity of theoretical background - documentation, theoretical explanations *

1 2 3 4 5 6 7 8 9 10

very low ○ ○ ○ ○ ○ ○ ○ ○ ○ ○ very high

Explain / give examples - background

Fig. 2 Example of an item from the LFQ questionnaire

in the educational literature [5, 22, 23], includes 81 items. We selected 31 items from the questionnaire in the following seven categories, which are particularly relevant to the context of engineering education:

- Intrinsic goal orientation
- Extrinsic goal orientation
- Orientation
- Task value
- Control of learning beliefs
- Self-efficacy for learning and performance
- Metacognitive self-regulation

Other studies in which only relevant sub-scales from the original questionnaire were used were presented by Bassili [24] and Al Khatib [25]. Artino [26], who examined in detail the history of the MSLQ and its use, writes that "this instrument is completely modular, and thus the scales can be used together or individually, depending on the needs of the researcher, instructor, or student" (p. 4).

In this case as well, the students were asked to explain or give examples of their answers in an open text box, as illustrated in the example from the MSLQ questionnaire shown in Fig. 3.

Category: *Intrinsic goal orientation*

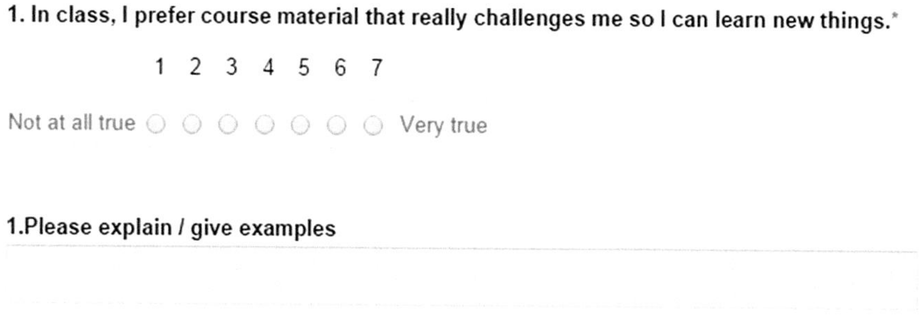

1. In class, I prefer course material that really challenges me so I can learn new things. *

1 2 3 4 5 6 7

Not at all true ○ ○ ○ ○ ○ ○ ○ Very true

1.Please explain / give examples

Fig. 3 Example of an item from the MSLQ questionnaire

4.4 The Computer System Usability Questionnaire (CSUQ)

The CSUQ is a tool for measuring user satisfaction in using computerized systems (hardware and software). The tool was developed at IBM for internal use by Lewis and others [27] and later on become an acceptable tool academically. Usability questionnaires gather subjective and objective data in realistic scenarios-of-use. Subjective data, for example, are measures of participants' opinions or attitudes concerning their perception of usability. Objective data are measures of participants' performance, such as scenario completion time and successful scenario completion rate.

The CSUQ consists of 19 items from four categories:

- Overall satisfaction
- System usability
- Information quality
- Interface quality

The full version of the questionnaire is presented in Appendix 2. The students were asked to mark their answers on a seven-point Likert scale from 1 (totally agree) to 7 (totally disagree), and also explain or give examples of their answers in a free text box, as illustrated in Fig. 4. Since the original scale mentioned above is defined in an unconventional order, it was inverted in the data analysis to represent the range from 1-totally disagree to 7-totally agree, as is frequently used in social science research. In the current study, the students filled in the CSUQ once per course after completing the lab work.

4.5 Qualitative Evaluation

In addition to the quantitative evaluation tools described above, the current study also involved qualitative tools, mainly:

Semi-structured interviews held with students and teachers
Observations carried out in class sessions

Analysis of students' comments in the open-ended text boxes in the LFQ, MSLQ and CSUQ questionnaires described above

3. The interface of this system is pleasant. *

Note: The interface includes those items that you use to interact with the system. For example, some components of the interface are the keyboard, the mouse, the screens (including their use of graphics and language).

1 2 3 4 5 6 7

strongly agree ○ ○ ○ ○ ○ ○ ○ strongly disagree

COMMENTS 3:

Fig. 4 Example of an item from the CSUQ questionnaire

The aim of the interviews was to shed light on the processes of teaching, learning, achievements and motivation, and expose cases of success or difficulties that cannot be expressed in the quantitative data. According to Schutt [28, p. 348], "Conducting qualitative interviews can often enhance the value of a research design that uses primarily quantitative measurement techniques." The interviews held with the teachers and students also aimed at collecting data about the technical quality of the E2LP boards, bugs in the hardware and software, and participants' suggestions about how to improve the system. The interviews with the teachers and students were conducted by a member of the evaluation team from another university in the consortium.

4.6 Data Analysis Method

The quantitative data obtained from the LFQ, MSLQ and CSUQ questionnaires were in the ordinal scale, for example, 1, 2, 3, 4, 5, 6, 7. In these data types, we can say that 2 is greater than 1 and 3 is greater than 2, but the numbers just represent an order and their specific values have no meaning. For example, the scale can also be represented by the letters a, b, c … Therefore, the reliability (internal consistency) of a ordinal data set is measured by the 'Ordinal Alpha' coefficient, which replaces the Cronbach's Alpha coefficient, often used as a measure of reliability for continuous data [29, 30]

Relative to the qualitative data, according to the literature, the data analysis tends to be inductive—the analyst identifies the important categories in the data, as well as patterns and relationships, through a process of discovery. There are often no predefined measures or hypotheses [28, 31, 32]. In the current case, we: (1) read all the material several times to identify the categories according to the participants' responses; (2) chose the main categories to be addressed in the Findings section; (3) searched the text again and marked all the cases in which the students related to these categories; (4) summarized the findings, as presented in the following sections; and finally (5) examined how the findings from the quantitative data relate to answers in the LFQ and SRL qualitative questionnaires.

5 Findings

5.1 Findings from the Lab Feedback Questionnaire (LFQ)

As previously noted, the LFQ was administrated five times during the course in order to receive feedback from the students in 'real' time. The questionnaire included the following categories:

- Clarity of theoretical background—documentation, theoretical explanations.
- Clarity of technical instructions, exercises and problems.
- Total time and efforts required.

- Ease of use of the environment—software issues.
- Ease of use of the platform—hardware.
- Feeling of immersion.
- To what extent do you think you learned something valuable?
- Overall satisfaction.

The students marked their answers on a scale of 1 (very low) to 10 (very high). In Figs. 5 and 6, we present only part of the findings in order to demonstrate how data were gathered during the course.

Figure 5 presents the students' answers to all eight categories in the LFQ in the last round (no. 5). It can be seen that the students marked quite positive answers to most of the categories in the questionnaire, except for category no. 3—total time and efforts required—which indicates that the students worked hard in this lab.

On the other hand, Fig. 6 presents the students' answers to category no. 8—'Overall satisfaction'—in rounds 1–5 of running the LFQ. It can be seen that the learners' overall satisfaction from the lab work decreased gradually from labs 1 to 4, and increased again in lab 5. This finding can be explained by the fact that the labs became increasingly more complex, on the one hand, but the students gained more experience and confidence in using the system, on the other hand.

5.2 Findings from the MSLQ Questionnaire

As noted earlier in the Method section, at the end of the semester the students filled in the Motivated Strategies for Learning Questionnaire (MSLQ) that measures aspects of motivation and metacognition according to the Self-regulated Learning

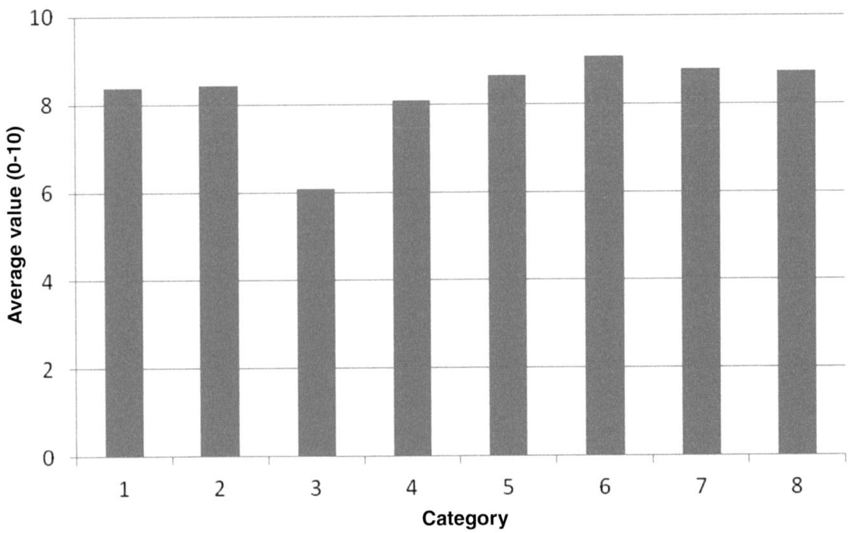

Fig. 5 Students' answers to all eight categories of the LFQ in the last round (no. 5) (n = 56)

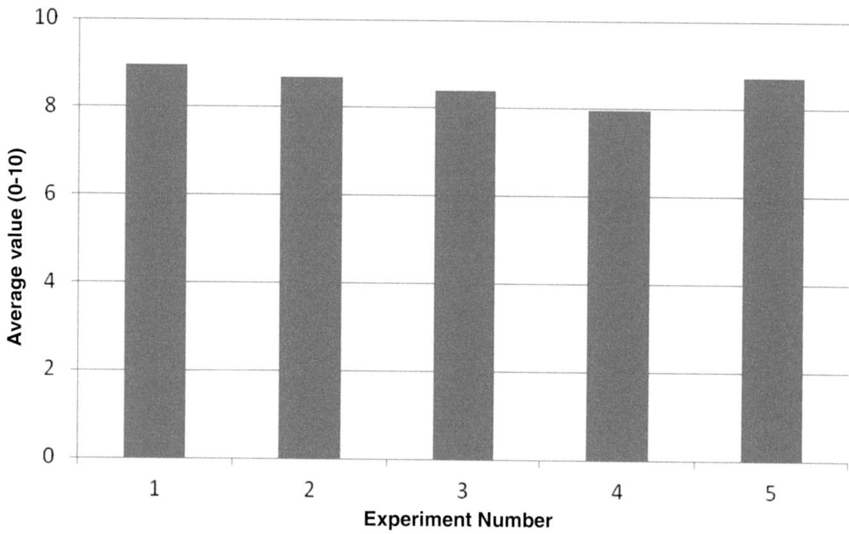

Fig. 6 Students' answers to Category no. 8 'Overall satisfaction' in rounds 1-5 of the LFQ

(SRL) theory. The findings from this questionnaire are presented in Table 1 and shown graphically in Fig. 7. As each sub-scale in the questionnaire is composed of a number of items, the reliability (internal consistently) of the findings was checked by calculating the Ordinal Alpha coefficient for each sub-scale, as shown in Table 1. As previously noted, this method of estimating reliability or ordinal data, such as answers on the Likert-type scale, replaces the well-known Cronbach's Alpha coefficient, which was originally developed for data on an interval scale or ratio scale. According to the literature [33, 34], the ranges of Alpha values are considered as follows: $\alpha \geq 0.9$ = excellent; $0.7 \leq \alpha < 0.9$ = good; $0.6 \leq \alpha < 0.7$ = acceptable; $0.5 \leq \alpha < 0.6$ = poor; $\alpha < 0.5$ = unacceptable. Consequently, the reliability of the MSLQ sub-scales displayed in Table 1 is "good."

The findings presented in Table 1 and Fig. 7 show that all of the students' answers to the six sub-scales in the MSLQ questionnaire are quite positive. The highest mean scores were obtained for the sub-scales "Control of learning beliefs" (for example, awareness of how I learn) and "Self-efficacy for learning" (for

Table 1 Findings from the MSLQ questionnaire for the Logic Design of Computer Systems 1 course (n = 111)

Sub-scale	Ordinal Alpha	Mean score	Std. Dev
Intrinsic goal orientation (4 items)	0.74	5.45	0.95
Extrinsic goal orientation (4 items)	0.74	4.35	1.37
Task value (6 items)	0.84	5.50	0.55
Control of learning beliefs (4 items)	0.96	5.75	0.93
Self-efficacy for learning and performance (8 items)	0.90	5.69	0.82
Meta-cognitive self-regulation (5 items)	0.75	4.96	1.06

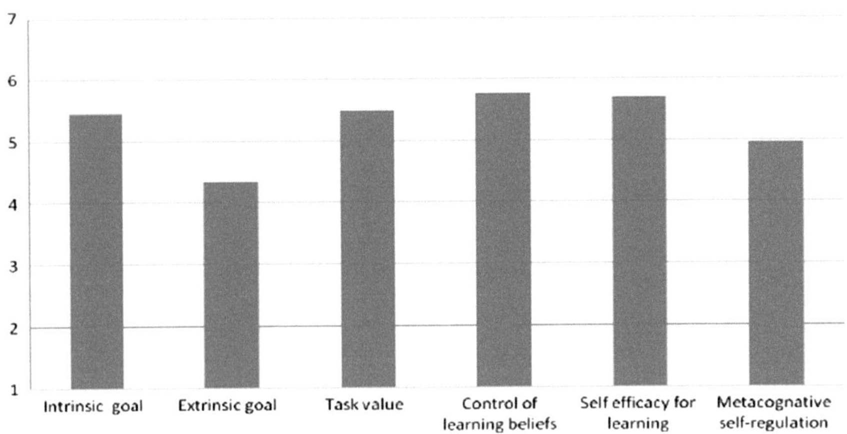

Fig. 7 Mean value of the answers to sub-scales of the MSLQ questionnaire for the Logic Design of Computer Systems 1 course

example, confidence in my abilities to succeed in this course). The lowest mean score was found for the category "Extrinsic goal orientation" (for example, course grades).

The findings from the MSLQ questionnaire described above indicate that the students expressed positive beliefs about their abilities to learn, they are mature enough to take responsibility for their learning, and their internal motivation drives them to learn more than external factors. The students not only grasped the importance of learning this subject for their professional career, but were also internally motivated to cope with the challenges this course presented to them. These findings are encouraging because they align with the educational literature showing that internal motivation (rather than external motivation) is one of the most important factors in people's learning and personal development [35].

5.3 Findings from the Computer System Usability Questionnaire (CSUQ)

As previously explained, the CSUQ questionnaire (shown in Appendix 2) is a tool for measuring students' satisfaction in using a computerized system, hardware and software. Since, in the current research, the students carried out all of the lab work in a computerized environment, it was particularly important to obtain learners' feedback in this regard. In addition, we asked the teachers to evaluate the computerized learning enviromant according to this questionnaire.

In this case as well, the Ordinal Alpha coefficient was calculated to measure the reliability of the data obtained in the four categories, as presented in Table 2.

Table 2 Reliability (internal consistency) of the findings from the CSUQ questionnaire

Category	Ordinal Alpha Students (n = 21)	Ordinal Alpha Teachers (n = 9)
Overall (2 items)	0.40	1.00
System use (8 items)	0.91	0.91
Information (7 items)	0.97	0.84
Interface quality (2 items)	0.66	0.59

It can be seen in Table 2 that the internal consistency (reliability) was low ($\alpha = 0.2$) for students' answers to the two items in the category 'Overall' and marginal ($\alpha = 0.59$) for teachers' answers to the two items in the category 'Interface quality.' This can be partially explained by the small samples of students and teachers who answered the questionnaire (only 10 % of the students who learned the course). Therefore, the average scores of students' answers to the items in the category 'Overall' are presented separately in Fig. 8.

The findings presented in Fig. 8 show that in general the students were reasonably satisfied with the E2LP platform and the software. The average lower score was marked for the category 'Information quality,' which had to do with insufficient or unclear information that was included in the system documentation and user's instructions. This point was also observed in the interviews held with the students, as reported in the following section.

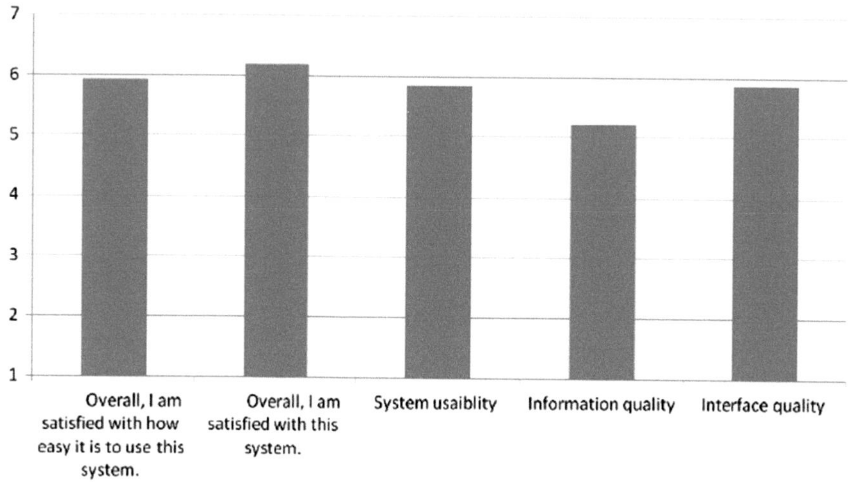

Fig. 8 Student's answers to the CSUQ questionnaire in the Logic Design of Computer Systems 1 course (scale 1–7)

5.4 Findings from Students' Comments in the Open Text Boxes in the Questionnaires

As previously mentioned, the LFQ, MSLQ and CSUQ questionnaires included open-ended text boxes in which the students could add comments and suggestions. About 15 % of the students (65 out of 433) added comments in the LFQ questionnaire, 11 % (12 out of 112) wrote comments in the MSLQ questionnaire, and 19 % (5 out of 26) wrote comments in the CSUQ questionnaire. In the data analysis, we identified three types of comments the students had: positive, critical and suggestive. Examples from the comments the students wrote in the LFQ questionnaire are shown below.

An example of a positive comment a student wrote in regard to Lab 1:

> Algorithm steps are clustered in a logical manner; very clear instructions and good usage of diagrams, as well as writing necessary steps like a list, instead of writing a single paragraph of text (which would reduce readability and maybe the clarity of the problem).
>
> Good use of the UML diagram, clear assignments that were very well segmented. Every step presented in a new sentence and with a bullet helps to get a grip on this task in order to proceed with it.

An example of a negative comment a student wrote in regard to Lab 4:

> Probably the most difficult part was to understand the exercise. This one required a lot of effort.
>
> The assistants explained in theory what we needed to do, but not how to do it. For example, the first time we were making one project out of the components (top), they told us that we needed to connect signals from all the components with signals in the top module, but they didn't tell us how to do this.
>
> Good, but not good enough, we still had tons of questions and needed more exercises.

An example of a suggestive comment a student wrote in regard to Lab 4:

> The 4th exercise is more difficult than the previous ones so I believe there should be an exercise in between the 3rd and the 4th.
>
> The task is pretty clear. However, I do mind that the task explicitly asks for designing a finite state machine; this problem could also be solved in an elegant manner using a 2-bit shift register.
>
> Even though the point of any class is to learn something on our own, it would be really nice to add some context about loops, which would shorten the time required somewhat.

It is worth mentioning that the findings reported in the paper are from the second year of running the program in the classes. Similar comments that had been received in the first year contributed to updating both the software and lab experiments developed in the project.

While students' comments in the LFQ questionnaire related to specific lab experiments, the comments they wrote in the MSLQ questionnaire dealt with more general aspects of learning and motivation, as seen in the following examples.

Example of students' open comments in the MSLQ questionnaire

About Intrinsic Goal Orientation

Because of my personal motivation and love for subjects and knowledge, I learn more.

About Task Value

Although I will probably not use this knowledge further in some extent of VHDL programming, the knowledge is fundamental for anybody working with computing systems or programming.

About Self-efficacy

I always like to believe that you can accomplish anything if you work hard enough.

These examples of comments the students had in the MSLQ questionnaire teach us that the participants understood the meaning of the subcategories in the questionnaire and addressed them seriously. These authentic data are valuable, because they complement the students' answers to the closed-ended items in the questionnaire and the points they raised in the face-to-face interviews.

Comments the students added to the CSUQ questionnaire

Only a few students added comments in the open-ended text boxes in the CSUQ questionnaire. The main issues were finding information and error messages. The teachers wrote, for example:

No online help; messages in software could be better" and "Error messages are still the issue, better than last year, but not perfect." About documentation, they wrote "Documentation needs improvement, more effort will be invested for next year (course).

5.5 *Findings from the Interviews with Students and Teachers*

Data analysis was done by closely reading the data obtained, identifying the main categories in the participants' responses, searching the text again and marking all the cases in which the students related to these categories, and examining how the findings from the quantitative data related to the participants' answers in the LFQ and SRL qualitative questionnaires.

Since we had gathered a large amount of data in this part of the evaluation, only part of it can be presented here.

Findings from interviews with the students and teachers

In the course under discussion, a member from the evaluation team interviewed 36 students in two focus groups (total 2.5 h) and 12 teachers in focus groups (1 h), and conducted an in-depth interview with one teacher. All the interviews were recorded and transcribed. Data analysis included three consecutive rounds:

Identifying three-four main categories that arose in the discussion
Identifying three-four subcategories of each main category
Identifying examples for each category

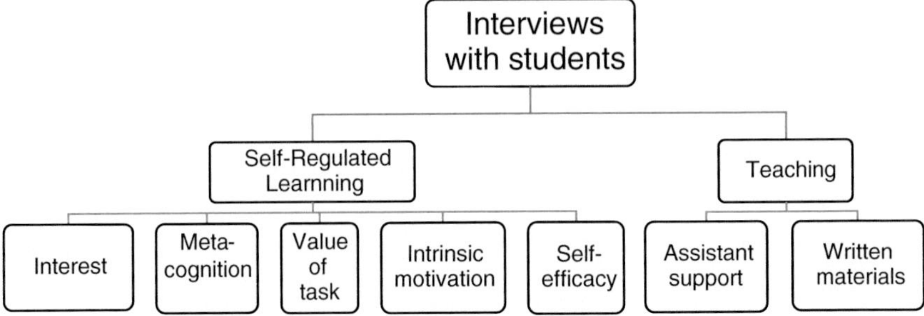

Fig. 9 Main categories and subcategories identified from interviews with students

The main categories and subcategories identified in the interviews with the students are illustrated in Fig. 9.

Below are some examples of comments the students had in the interviews according the categories shown in Fig. 9:

Teaching > Assistant support

> We had problems with one of the assistants who went out all the time to smoke and didn't help us.

Self-Regulated Learning > Self-efficacy

> Usually I'm confident at the beginning of the semester but I became less confident during the semester.

Self-Regulated Learning > Intrinsic Motivation

> I think I like courses that challenge me, because if you deal with material that demands using your intelligence it improves you. You improve your skills and grades.

Self-Regulated Learning > Value of Task

> I think that what we learned in the DSD 1 course is unique to this subject and I don't see how it is applicable to other subjects, but it changed my way of thinking. For example, when I walk in the street and see an LCD commercial I now know how it works.

Findings from the interviews with the course teachers and assistants

Figure 10 illustrates the categories and subcategories that were identified in analyzing the data obtained from the interviews with the teachers and lab assistants,

Below are some comments by the participants during the interviews:

Pedagogy > Teaching

> There is improvement in the frontal lessons during which we practiced the experiments on paper. I synced the labs with the lectures and spent most of the time practicing the experiments.

Pedagogy > Course Curriculum

> Last year when projects were mandatory we had complaints from the students that LPRS1 (the course) was very demanding. By making projects optional, we allow the students to divide into two groups—those who just want to pass the course and those who want to deepen their knowledge.

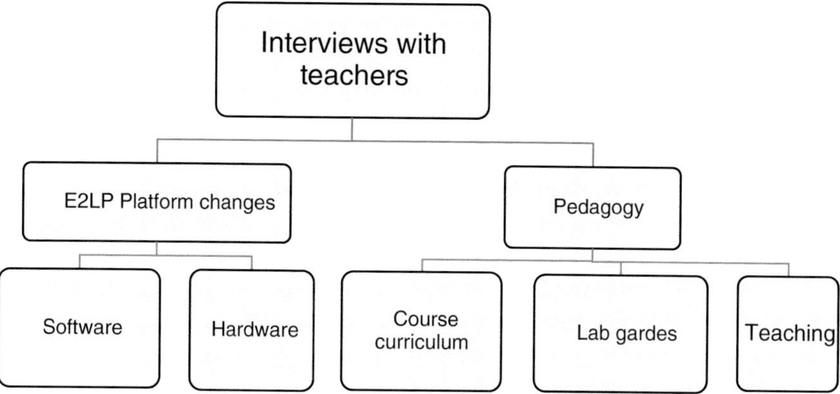

Fig. 10 Main categories and subcategories that were identified from the interviews with the teachers and lab assistants

These examples from the interviews with the teachers show that introducing the new platform into the course program and participating in the evaluation program caused them to reflect on the course curriculum and instruction methodology.

6 Discussion and Conclusions

This study addressed the case of the development, implementation and evaluation of an innovative platform for teaching computer and embedded systems engineering. The platform under development included hardware, software and instructional materials for students. The lab experiments were designed on three levels—exercises, projects and projects—according to the 'Task Taxonomy' defined in this study. The evaluation program adopted the internal-formative evaluation approach that emphasizes close work with the program developers and the teachers in the classes [3]. The evaluation was derived from the Self-Regulated Learning Theory [8, 11], which relates to cognitive, meta-cognitive, motivational and self-efficacy aspects of learning. Data were collected by administrating the Lab Feedback Questionnaire (LFQ), the Motivated Strategies for Learning Questionnaire (MSLQ), the Computer System Usability Questionnaire (CSUQ), and interviews held with students and teachers.

The current study aimed at identifying elements of the learning platform that helped or hindered students' success and motivation in learning the courses, and shed light on the role the evaluation methodology played in program development and implementation.

Although the current study did not deal with measuring students' course grades, it is worth examining students' scores in the labs, as shown in Fig. 11.

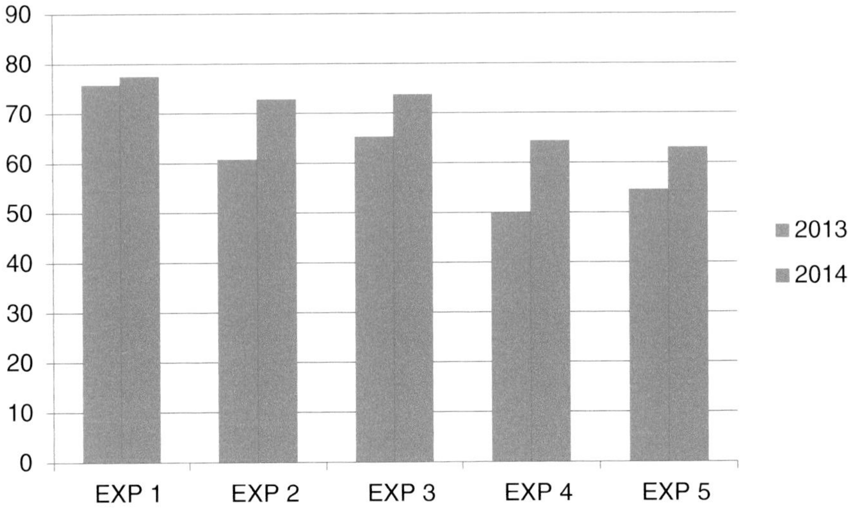

Fig. 11 Students' scores in five lab experiments in the years 2013 (n = 224) and 2014 (n = 263)

It is important to mention that the students' work in the labs took place in small groups under the instruction of nine teachers, who also gave the scores. Figure 11 shows two interesting points. First, it can be seen that the students' scores in the labs decreased as they advanced from one experiment to another. This reflects the fact that the experiments' complexity increased from stage to stage—from exercises, to problems and projects—as the students also noted in the Lab Feedback Questionnaire (LFQ). Second, Fig. 11 shows that students' scores in all five experiments increased in the year 2014 in comparison to the previous year, when the E2LP platform was used for the first time in the lab. This change could be attributed to improving the system software, updating the instructional materials, and the fact that the teachers gained more experience in using the system. The findings from the evaluation process and the close work of the evaluation team with the developers and the teachers certainly contributed to these outcomes.

In summary, there is full agreement that in developing an advanced computer-based learning system, the hardware and software should work properly and be adapted to a wide range of users, from beginners to experts. Beyond that, the experience accumulated in the current research taught us that one of the most important factors affecting the success of an advanced technological learning platform is the careful design of students' assignments, for example, to progress gradually from solving basic exercises, to solving more significant problems and dealing with broad projects.

With regard to the evaluation methodology used in this research, one of the advantages of the applied method was addressing a number of aspects of learning and motivation, and combining qualitative and qualitative methodologies and tools. More specifically, the short Lab Feedback Questionnaire (LFQ) the students answered five times during the course, including writing comments in an open text box, as well as the interviews held with the students and teachers, had the greatest contribution to the effectiveness of the evaluation and the project's success. Among the important points observed during the evaluation process by the MSLQ questionnaire was the potential of the new learning platform in fostering students' intrinsic motivation to deal with challenging tasks in the computer engineering lab.

Appendix 1

Lab Feedback Questionnaire (LFQ)

Dear Student,

Pleas mark the **effectiveness** of the lab experiment on the scale **1 … 10**

1. **Clarity of theoretical background**—documentation, theoretical explanations

   (1-very low … 10-very high) ___ Explain/give example:

2. **Clarity of technical instructions, exercises and problems**

   (1-very low … 10-very high) ____ Explain/give example:

3. **Total time and efforts required**

   (1-very much … 10-very little) ____ Explain/give example:

4. **Ease of use—the environment,** Xilinx software and BIN download software

   (1-very difficult …10-very easy) ____ Explain/give example:

5. **Ease of use—the platform**

   (1-very difficult …10-very easy) ____ Explain/give example:

6. **Feeling of immersion**—being part of the environment, control over the system

   (1-very low … 10-very high) ___ Explain/give example:

7. **To what extent do you think you learned something valuable**?

   (1-very low … 10-very high) ___ Explain/give example:

8. **Overall satisfaction**

   (1-very low … 10-very high) ___ Explain/give example:

Appendix 2

Computer System Usability Questionnaire (CSUQ)

The CSUQ consists of 19 items from four categories:

a. Overall satisfaction (items 1, 19)
b. System usability (items 2, 4.6, 7, 9, 11, 17, 18)
c. Information quality (items 5, 8, 10, 12, 13, 15, 16)
d. Interface quality (items 3, 14)

Scale for students' answers, to each item (NA—Not Applicable)

Strongly AGREE	1	2	3	4	5	6	7	NA	Strongly DISAGREE

Comments:

____

____

1. Overall, I am satisfied with how easy it is to use this system.
2. It is simple to use this system.
3. The interface of this system is pleasant.
4. I am able to complete my work quickly using this system.
5. It is easy to find the information I need.
6. I am able to efficiently complete my work using this system.
7. I feel comfortable using this system.
8. The information (such as online help, on-screen messages and other documentation) provided with this system is clear.
9. I believe I became productive quickly using this system.
10. The system gives error messages that clearly tell me how to fix problems.
11. I can effectively complete my work using this system.
12. Whenever I make a mistake using the system, I recover easily and quickly.
13. The information provided with the system is easy to understand.
14. I like using the interface of this system.
15. The system information (such as online help, on-screen messages and other documentation) is effective in helping me complete my work.
16. The organization of information on the system screens is clear.
17. This system has all the functions and capabilities I expect it to have.
18. It was easy to learn to use this system.
19. Overall, I am satisfied with this system.

References

1. Kastelan, I., Lopez, B.J.R., Artetxe, G.E., Piwinski, J., Barak, M., Temerinac, M.: E2LP: a unified embedded engineering learning platform. Microprocess. Microsyst. **38**(8), 933–946 (2014)
2. Zagar, M., Frid, N., Knezovic, J., Hofman, D., Kovac, M., Sruk, V., Mlinaric, H.: Unified, multiple target, computer engineering learning platform: design results and learning outcomes. In: IEEE Global Engineering Education Conference, EDUCON, pp. 926–929 (2014)
3. Owens, J.M.: Program Evaluation. The Guilford Press, New York (2007)
4. Boekaerts, M.: Self-regulated learning: where we are today. Int. J. Educ. Res. **31**(6), 445–457 (1999)
5. Pintrich, P.R.: Educational Resources Information Center (U.S.). A Manual for the Use of the Motivated Strategies for Learning Questionnaire (MSLQ). University of Michigan, Ann Arbor, Mich (1991)
6. Zimmerman, B.J.: Self-regulated learning and academic achievement: an overview. Educ. Psychol. **25**, 3–17 (1990)
7. Flavell, J.H.: Metacognition and cognitive monitoring: a new area of cognitive-developmental inquiry. Am. Psychol. **34**(10), 906–911 (1979)
8. Pintrich, P.R.: A conceptual framework for assessing motivation and self-regulated learning in college students. Educ. Psychol. Rev. **16**(4), 385–407 (2004)
9. Schon, D.A.: Educating the Reflective Practitioner: Toward a New Design for Teaching and Learning in the Professions. Jossey-Bass Inc, San Francisco (1996)
10. Deci, E.L., Ryan, R.M.: Facilitating optimal motivation and psychological well-being across life's domains. Can Psychol **49**(1), 14–23 (2008)
11. Bandura, A.: Self-Efficacy: The Exercise of Control. WH Freeman and Company, New York (1997)
12. Barak, M.: Motivating self-regulated learning in technology education. Int. J. Technol. Des. Educ. **20**(4), 381–401 (2010)
13. Lawanto, O., Santoso, H.B., Goodridge, W., Butler, D., Cartier, S.: Task interpretation, cognitive, and metacognitive strategies of higher and lower performers in an engineering design project: an exploratory study of college freshmen. Int. J. Eng. Educ. **29**(2), 459–475 (2013)
14. Schraw, G., Kent, J.C., Kendall, H.: Promoting self-regulation in science education: metacognition as part of a broader perspective on learning. Res. Sci. Educ. **36**(1–2), 111–139 (2006)
15. Thomas, J.W.: A Review of Research on Project-Based Learning, Autodesk, San Rafael, CA. (2000). Retrieved March 15, 2009, from http://www.bie.org/files/researchreviewPBL.pdf
16. Barak, M., Shachar, A.: Project in Technology and Fostering Learning: The Potential and its Realization. J. Sci. Educ. Technol. **17**(3), 285–296 (2008)
17. Kirschner, P.A., Sweller, J., Clark, R.E.: Why minimal guidance during instruction does not work: an analysis of the failure of constructivist, discovery, problem-based, experiential, and inquiry-based teaching. Educ. Psychol. **41**(2), 75–86 (2006)
18. Barak, M.: Teaching engineering and technology: cognitive, knowledge and problem-solving taxonomies. J Eng Des Technol **11**(3), 316–333 (2013)
19. Olds, B.M., Moskal, B.M., Miller, R.L.: Assessment in engineering education: evolution, approaches and future collaborations. J. Eng. Educ. **94**(1), 13–25 (2005)
20. Borrego, M., Douglas, E.P., Amelink, C.T.: Quantitative, qualitative, and mixed research methods in engineering education. J. Eng. Educ. **98**(1), 53–66 (2009)
21. Koro-Ljungberg, M., Douglas, E.P.: State of qualitative research in engineering education: meta-analysis of JEE Articles, 2005–2006. J. Eng. Educ. **97**(2), 163–175 (2008)
22. Lawanto, O., Santoso, H.B., Yang, L.: Understanding of the relationship between interest and expectancy for success in engineering design activity in Grades 9–12. J. Educ. Technol. Soc. **15**(1), 152–161 (2012)

23. Rotgans, J.I., Schmidt, H.G.: The motivated strategies for learning questionnaire: a measure for students' general motivational beliefs and learning strategies? Asia-pacific Educ. Res. **19**(2), 357–369 (2010)
24. Bassili, J.N.: Motivation and cognitive strategies in the choice to attend lectures or watch them online. J. Distance Educ. **22**(3), 129–148 (2012)
25. Al Khatib, S.A.: Meta-cognitive self-regulated learning and motivational beliefs as predictors of college students' performance. Int. J. Res. Educ. **27**, 57–72 (2010)
26. Artino, A.R.J.: Review of the Motivated Strategies for Learning Questionnaire. University of Connecticut, ERIC Number: ED499083. (2005). http://eric.ed.gov/?q=artino&pg=2&id=ED499083
27. Lewis, J.R.: IBM computer usability satisfaction questionnaires: psychometric evaluation and instructions for use. Int. J. Hum.-Comput. Interact. **7**(1), 57–78 (1995)
28. Schutt, R.K.: Investigating the Social World: The Process and Practice of Research. Sage Publications, Thousand Oaks (2012)
29. Gadermann, A.M., Guhn, M., Zumbo, B.D.: Estimating ordinal reliability for likert-type and ordinal item response data: a conceptual, empirical, and practical quide. Pract. Assess. Res. Eval. **17**(3), 1–13 (2012)
30. Zumbo, B.D., Gadermann, A.M., Zeisser, C.: Ordinal versions of coefficient alpha and theta for likert rating scales. J. Modern Appl. Stat. Methods **6**(1), 21–29 (2007). http://digitalcommons.wayne.edu/jmasm/vol6/iss1/4
31. Denzin, N.K., Lincoln, Y.S.: Handbook of Qualitative Research. Sage Publications, Thousand Oaks (2000)
32. Patton, M.: Qualitative Research and Evaluation Methods. Sage Publications, Thousand Oaks (2002)
33. Kline, P.: The Handbook of Psychological Testing, 2nd edn. Routledge, London (2000)
34. Nunnaly, J.: Psychometric Theory. McGraw-Hill, New York (1978)
35. Deci, E.L., Ryan, R.M.: Intrinsic Motivation and Self-Determination in Human behavior. Plenum Press, New York (1985)

Is It Possible to Increase Motivation for Study Among Sophomore Electrical and Computer Engineering Students?

Aharon Gero

Abstract In the Department of Electrical Engineering at the Technion—Israel Institute of Technology—a unique course is being held—"Introductory Project in Electrical Engineering". The course is intended to expose second-year students of electrical and computer engineering to the occupational areas of electrical and computer engineers, to strengthen their sense of relatedness to the Department and increase their intrinsic motivation towards their studies. The core of the course is a team project for planning a window-cleaning robot. Using quantitative and qualitative tools, the study described in this paper has followed students who participated in the course, examining changes in the motivational factors driving them to study electrical and computer engineering. Findings point out a significant improvement in students' intrinsic motivation and identified regulation—an improvement explained in light of the self-determination theory.

Keywords Electrical and computer engineering education · Introductory engineering course · Motivation

1 Introduction

The course "Introductory Project in Electrical Engineering" is an elective course, offered by the Department of Electrical Engineering at the Technion—Israel Institute of Technology. It is intended for second-year electrical and computer engineering students. The main purpose of the course is to expose students to the occupational areas of electrical and computer engineers and to strengthen their sense of relatedness to the Department in order to increase their motivation.

A. Gero (✉)
Technion—Israel Institute of Technology, 32000 Haifa, Israel
e-mail: gero@technion.ac.il
URL: http://edu.technion.ac.il/en/gero

R. Szewczyk et al. (eds.), *Embedded Engineering Education*,
Advances in Intelligent Systems and Computing 421,
DOI 10.1007/978-3-319-27540-6_11

A secondary purpose that the course is meant to achieve is improving the systems thinking of the students—a way of thinking which focuses on understanding the interactions between the various system components, and the resulting synergy [1]. The core of the course is a project for planning a window-cleaning robot, which is being carried out in teams.

The idea behind the development of the course was that similarly to other universities, students of the Department of Electrical Engineering mostly study mathematics and physics in their first three semesters of study—courses delivered by the teaching staff of those departments. This fact contributes to a feeling of detachment among the students, both from the discipline of electrical and computer engineering, and from the Department teaching staff [2].

Introductory courses with similar purposes are offered by universities to electrical and computer engineering students [3, 4], mechanical engineering students [5, 6] or jointly to all engineering students at the institution [7]. It is reported in literature that these introductory courses have managed to improve students' understanding concerning the fields included in the discipline [2], and to increase their motivation to study [7].

The study described hereunder examined changes that occurred in students' motivation following their participation in the course. Changes which occurred during the course in the systems thinking skills of the students—are described in [8]. The studies focusing in introductory courses, which have been quoted above, treated motivation as a single entity. In the present study, we distinguish between the various motivational factors, in view of the definitions of the self-determination theory [9, 10], and thus refine our findings. Furthermore, unlike the above studies, which have taken a constructivist-qualitative approach [11], the study discussed is a mixed-method study, using quantitative as well as qualitative tools, in order to present various aspects of the investigated phenomenon, and increase the trustworthiness of its findings [12].

The paper begins with theoretical background reviewing the self-determination theory. Later, the course "Introductory Project in Electrical Engineering" is described, and the study objective and the chosen methodology are presented. After the description of the findings, a discussion is conducted.

2 Theoretical Background

Motivation is defined as the individual's will to invest resources in a certain behavior. The source of motivation is explained through a wide array of theoretical approaches [13–15]. The self-determination theory [9, 10], being part of the humanistic approach, maintains that a person has three innate needs:

- The need for autonomy—the need to feel that the individual's behavior was not forced upon his/her, but emanates from his/her needs;
- The need for competence—the need to feel that the individual is capable, and can meet challenging objectives;

- The need for relatedness—the need to love and be loved, and to be a part of a group.

Where a person's needs are satisfied, this will bring him/her to a high level of motivation, while the prevention of this satisfaction will compromise it. The self-determination theory describes the origin of motivation as a spectrum spanning between extrinsic and intrinsic factors. At one end lies Extrinsic Motivation, which includes four types of regulation:

- External regulation—emanating from hopes of receiving material remuneration, or alternatively, from fear of punishment;
- Introjected regulation—caused by considerations of personal prestige or by the desire to fulfill expectations of people who are important to the individual;
- Identified regulation—stemming from identifying the value of the behavior. The behavior is a means allowing another activity which causes interest and pleasure, or alternatively, the behavior carries moral value;
- Integrated regulation—emanating from viewing the behavior as reflecting the identity of the individual.

At the opposite end lies the Intrinsic Motivation, emanating from interest and pleasure. In addition, the theory defines a situation of Amotivation, where the individual is bereft of any intention to act, because he/she attributes lack of importance to the activity, or lack of any feeling of competence to do it. The theory states that as much as motivation emanates from more intrinsic factors, the higher its quality will be.

The self-determination theory in the educational context [16] serves as a theoretical framework for many studies engaging in motivation among university students [17, 18]. Thus, for example, Koh et al. [19] have shown that simulation-based learning among mechanical engineering students provided the three needs mentioned above, and thus improved their intrinsic motivation. Similarly, Gero [20] has demonstrated that educational activity answering these needs among engineering students, has enhanced their intrinsic motivation. Since this theory has recently become a leading theory in the field of motivation in general, and of educational motivation in particular, we made use of it in this study.

3 Course Description

The semester course "Introductory Project in Electrical Engineering" includes one weekly two-hour meeting, and it is composed of two parts. The first seven weeks of the course include lectures and instruction on the nature of engineering and the differences between it and science (week 1), searching in online databases and creating an effective presentation (week 2), the occupational areas of electrical and computer engineers (weeks 3–4) and the engineering approach to problem solving (weeks 5–7). These contents provide students with the tools they will need to use during the second part of the course (weeks 8–14), focusing on planning a window-cleaning robot.

The project part of the course begins with an introductory lecture on robotics, and a presentation of the planning stages, week by week. Each week deals with a focused subject as detailed below: defining the robot structure and manner of movement (week 8), physical design (week 9), block diagram (week 10), integration of sensors (week 11), microcontrollers and drivers (week 12) and navigation algorithms (week 13). Each stage begins with a review of the design subject at hand, and upon completion, the students receive an assignment to complete in five-student teams, guided by a mentor who is a senior engineer at the Department. Thus, in week 8 the students are requested to collect information on window-cleaning robots, examine alternatives and decide upon the chosen solution; in week 9 they should choose energy sources and motors; in week 10, draw a block diagram of the robot; in weeks 11–12, select sensors, micro-controllers, and drivers; and finally, in week 13, prepare the final presentation, which is presented on the last week (week 14) before the staff and course participants. In the construction of the course, the following books were used: "Creative Problem Solving and Engineering Design" [21] and "Thinking Like an Engineer: An Active Learning Approach" [22].

It is important to note that the division of the design task to weekly sub-assignments was intended, according to the self-determination theory, to meet the students' need for competence. It was made clear to the students that the mentor is at their disposal, and that they can turn to him and consult him directly, but it is they who make their own design decisions, out of recognition of their independence and ability to make such decisions already at this stage of their studies. This method of work is intended to meet the students' relatedness and autonomy needs.

4 Research Goal

The aim of the study was to track changes in motivation towards electrical and computer engineering studies among students who participated in the "Introductory Project in Electrical Engineering" course.

The following research question was derived: Did a change occur in students' motivation to study electrical and computer engineering as a result of the course? If so—what characterizes this change?

5 Methodology

5.1 Participants and Method

Twenty-five students at the third semester of their undergraduate studies in electrical and computer engineering, who chose to participate in the "Introductory Project in Electrical Engineering" course, took part in the study. These students, who composed the experimental group, were requested to fill out an anonymous questionnaire at the beginning and at the end of the course, intended to qualify their motivational factors. Also, at the end of the course, five semi-structured interviews were conducted with the students, focusing on the students' attitudes towards electrical and computer engineering studies and towards the course. Throughout the course, observations took place at the study class. These observations focused on the behavioral aspect of the students, as it was expressed during the course.

In addition, thirty students at the third semester of their undergraduate studies in electrical and computer engineering, who did not participate in the discussed course, took part in the study. These students, who served as a control group, were requested to fill out the questionnaire at the beginning of the semester and at its end.

The quantitative findings were statistically analyzed, and the corresponding effect sizes were calculated. Through content analysis, which was based on the self-determination theory, the qualitative findings were sorted into categories. Only information coming up at least three times in the various research tools was included in this analysis.

5.2 Instruments

The questionnaire for characterization of motivational factors is a Likert-like five-level questionnaire, based on the SIMS (Situational Motivation Scale) questionnaire [23] and on the SRQ-A (Self-Regulation Questionnaire-Academic) questionnaire [24]. The questionnaire includes twenty statements reflecting four of the motivational factors mentioned in the theoretical part: intrinsic motivation, identified regulation, introjected regulation and external regulation. Similar to Gauy et al. [23], in order to keep the questionnaire from becoming too long, in a manner that will aggravate the respondents, we did not include statements in the questionnaire describing integrated regulation.

For example, the statement "I study electrical and computer engineering because I think it is interesting" reflects intrinsic motivation; the statement "I study electrical and computer engineering because I will benefit from it" reflects identified regulation; the statement "I study electrical and computer engineering because my parents want me to study it", as well as the statement "I study electrical and computer engineering because I want people to think I am smart" reflect introjected regulation; and the statement "I study electrical and computer engineering because I am supposed to do it" reflects external regulation. The statements were validated by two engineering education experts. Cronbach's alphas for each one of the motivational factors are: 0.84 (intrinsic motivation), 0.80 (identified regulation), 0.78 (introjected regulation) and 0.86 (external regulation). These values indicate a good level of internal consistency.

6 Findings

Figure 1 shows the mean score (which ranges between 20 and 100) given by the members of the experimental group to each one of the four motivational factors. Scores were given in the pretest questionnaire, filled out at the beginning of the course, and in the posttest questionnaire, filled out at its end.

An examination of the figure suggests that both in the beginning of the course and at its end, the intrinsic motivation score was the highest among the motivational factors, the identified regulation score was ranked second, the introjected regulation score was ranked third, and the external regulation score was the lowest. Additionally, it turned out that following the course, there was an increase in all motivational factors, with the exception of the external regulation factor, which has decreased.

Table 1 lists the scores (mean *M* and the standard deviation *SD*) given by members of both experimental and control groups to the various motivational factors. Following the *t*-test, there was no significant difference between the pretest questionnaire score of the experimental group and the pretest questionnaire score of the control group, regarding the four motivational factors. However, for intrinsic motivation and identified regulation, there was a significant difference between the posttest questionnaire score of the experimental group and the posttest questionnaire score of the control group. For introjected regulation and external regulation no significant difference was observed between the posttest questionnaire score of the experimental group and the posttest questionnaire score of the control group.

Table 2 shows the corresponding effect size for the various motivational factors. Examination of the table gives the impression that the gap in intrinsic motivation is characterized by a large effect size; the gaps in identified regulation and introjected regulation are accompanied by a medium effect size, and the (negative) gap in external motivation is negligible.

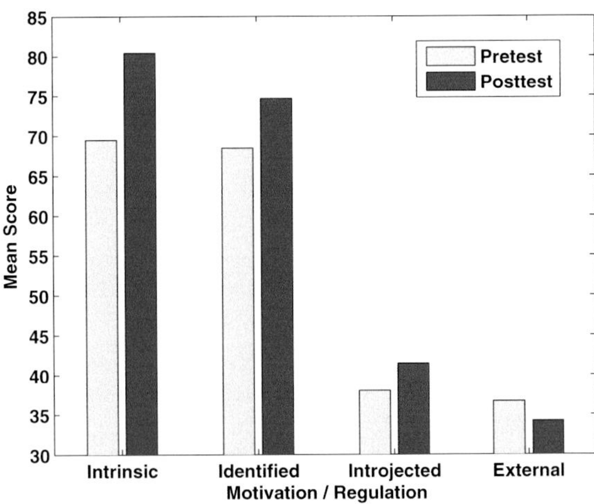

Fig. 1 Mean motivational factor score (experimental group)

Table 1 Motivational factor score (mean *M* and standard deviation *SD*)

Motivation	Regulation	Test	Group	*M*	*SD*	*p*-value
Intrinsic		Pretest	Experimental	69.50	10.53	n.s.
			Control	73.40	10.07	
		Posttest	Experimental	80.43	8.91	<0.01
			Control	69.58	17.13	
Extrinsic	Identified	Pretest	Experimental	68.48	10.62	n.s.
			Control	72.34	10.09	
		Posttest	Experimental	74.66	8.23	<0.01
			Control	67.06	15.08	
	Introjected	Pretest	Experimental	38.00	11.33	n.s.
			Control	35.33	10.63	
		Posttest	Experimental	41.45	13.74	n.s.
			Control	35.74	12.44	
	External	Pretest	Experimental	36.67	18.84	n.s.
			Control	33.07	9.12	
		Posttest	Experimental	34.20	15.87	n.s.
			Control	35.93	18.12	

Table 2 Effect sizes

Motivation	Regulation	Cohen's *d*
Intrinsic		0.78
Extrinsic	Identified	0.61
	Introjected	0.44
	External	−0.10

Table 3 Motivational factors (experimental group, end of experiment)

Motivational factor	Example	Interpretation
Intrinsic motivation	"My motivation increased because I was exposed to new, interesting areas that an electrical and computer engineer deals with."	Exposure to the occupational areas of electrical and computer engineers led to the creation of interest among the students
Identified regulation	"The course exposed me to the multiple employment options of an electrical and computer engineer."	Following exposure to the various occupational areas of electrical and computer engineers, the students recognized the high occupational value of the profession

Content analysis of the findings arising from the interviews (Table 3), makes it possible to explain the significant improvement in intrinsic motivation and identified regulation among the members of the experimental group, in that exposure to the various occupational areas of electrical and computer engineers has created interest and even recognition of the high occupational value of the profession.

Table 4 Satisfaction of needs (experimental group, end of experiment)

Need	Example	Interpretation
Autonomy	"We went totally wild, there was a great feeling… you decide and determine what the robot will look like and nobody decides for you."	The need for autonomy was satisfied thanks to the independence given to students during the project
Competence	"The process [of planning the robot] was guided with small assignments… they [course faculty] didn't throw us into the water and said 'good luck'… I felt that I can succeed in the project because they [course faculty] have divided our task into stages."	The need for competence was met by guidance and a focused definition of sub-assignment throughout all project stages
Relatedness	"Unlike the other courses in the Technion, where there is no one to talk to, I felt that here they [course faculty] were considerate and attentive to me, this is a difference of magnitude… I have a mentor whose door is always open…Now I totally feel that I'm a part of the Department."	The need for relatedness was satisfied thanks to the personal attitude of the course faculty

Support for the great interest created by the course emerges from the observations: throughout the course, the students took care to arrive on time, both in the beginning of the lesson and at the end of the break. This finding is far from reflecting the lateness culture, typical of Technion students, especially in courses taking place in the winter semester and starting at 8:30, such as the course discussed. Additionally, during the lessons, the students showed interest which was expressed, despite the early hour, in lively participation in the lectures on the one hand, and in not leaving the classroom for phone calls on the other hands, as happens very frequently in other courses.

On a deeper level, the interest created by the course among the students may be attributed to the satisfaction of their three fundamental needs, as detailed in Table 4.

7 Discussion and Outlook

The results of the study point to a significant improvement in intrinsic motivation (large effect size) and in identified regulation (medium effect size) among students who participated in the course. This improvement may be attributed to the students' exposure to the interesting occupational areas of electrical and computer engineering, and to the high occupational value of the profession. The improvement in intrinsic motivation carries great importance, as undergraduate studies in electrical and computer engineering require the development of higher-order thinking skills, and intrinsic motivation has a central role in this level of learning [25].

On a deeper level, the students' high level of motivation at the end of the course may be explained by the self-determination theory [9, 10], according to which the satisfaction of the individual's needs increases motivation. The need for autonomy was satisfied thanks to the independence granted to the students during the project; the need for competence was realized by the focused guidance and definition of sub-assignments throughout the project stages [26]; and finally, the need for relatedness was satisfied thanks to the personal attitude of the course faculty. These results match the findings presented in [19, 20], according to which a learning environment providing for the three needs under consideration, had improved the intrinsic motivation of students.

The study has two main limitations: a relatively small sample and a non-random assignment into experimental and control groups. In order to address the first limitation, qualitative tools, such as interviews and observations, were also used, in order to strengthen the trustworthiness of the findings. In order to address the second limitation, which is a regular characteristic of studies taking place in educational institutions, a preliminary test was performed, in order to rule out a significant difference between the two groups.

The theoretical significance of the study is in characterization, for the first time to the best of our knowledge, of the various motivational factors in studying electrical and computer engineering at early stages. The practical contribution may be expressed in the implications of the study conclusions upon planning curricula for engineers in general and electrical and computer engineers in particular.

In a continued study, we intend to find out whether the gaps found in the current study, between the scores of the experimental group and the scores of the control group, will also be preserved in the advanced years of study of the students. In addition, we would like to examine, upon completion of their studies, what the contribution of the course was to the qualification of the students as engineers.

Acknowledgments The author wishes to express his thanks to Nimrod Peleg, Kobi Kohai, and Avinoam Kolodny from the Department of Electrical Engineering at the Technion—Israel Institute of Technology—for their assistance.

This paper is based on the following paper: Gero, A.: Improving Intrinsic Motivation Among Sophomore Electrical Engineering Students by an Introductory Project. Int. J. Eng. Pedag. 2, 13–17 (2012)

References

1. Senge, M.: The Fifth Discipline: The Art and Practice of the Learning Organization. Doubleday, New-York (1990)
2. Pierre, J.W., Tuffner, F.K., Anderson, J.R., Whitman, D.L., Sadrul Ula, A.H.M.S., Kubichek, R.F., Wright, C.H.G., Barrett, S.F., Cupal, J.J., Hamann, J.C.: A one-credit hands-on introductory course in electrical and computer engineering using a variety of topic modules. IEEE Trans. Educ. **52**, 263–272 (2009)
3. Heer, D., Traylor, R.L., Thompson, T., Fiez, T.S.: Enhancing the freshman and sophomore ECE student experience using a platform for learning. IEEE Trans. Educ. **46**, 434–443 (2003)

4. Melton, R.W.: A Laboratory Approach to Multidisciplinary Freshman Computer Engineering. ASEE St. Lawrence Section Conference. New-York (2006)
5. Elata, D., Garaway, I.: A creative introduction to mechanical engineering. Int. J. Eng. Educ. **18**, 566–575 (2002)
6. Frank, M., Elata, D.: Developing the capacity for engineering systems thinking (CEST) of freshman engineering students. J. Syst. Eng. **8**, 187–195 (2005)
7. George, L.E., Brown, R.B.: Engineering 100: An Introduction to Engineering Systems at the US Air Force Academy. In: ASEE Annual conference, Honolulu (2007)
8. Gero, A.: Enhancing systems thinking skills of sophomore students: an introductory project in electrical engineering. Int. J. Eng. Educ. **30**, 738–745 (2014)
9. Deci, E.L., Ryan, R.M.: Intrinsic Motivation and Self Determination in Human Behavior. Plenum Publishing Co., New-York (1985)
10. Deci, E.L., Ryan, R.M.: The 'what' and 'why' of goal pursuits: human needs and the self determination of behavior. Psychol. Inq. **11**, 227–268 (2000)
11. Patton, M.Q.: Qualitative Evaluation Methods. Sage Publications, Beverley Hills (1980)
12. Keeves, J.P.: The unity of educational research. Interchange **19**, 14–30 (1988)
13. Bandura, A.: Self-efficacy: toward a unifying theory of behavioral change. Psychol. Rev. **84**, 191–215 (1997)
14. Weiner, B.: An Attribution Theory of Motivation and Emotion. Springer, Berlin (1986)
15. Ames, C.A.: Motivation: what teachers need to know. Teach. Coll. Rec. **91**, 409–421 (1990)
16. Deci, E.L., Vallerand, R.J., Pelletier, L.G., Ryan, R.M.: Motivation and education: the self-determination perspective. Educ. Psychol. **26**, 325–346 (1991)
17. Levesque, C., Zuehlke, A.N., Stanek, L.R., Ryan, R.M.: Autonomy and competence in German and American university students: a comparative study based on self-determination theory. J. Educ. Psychol. **96**, 68–84 (1991)
18. Kusurkar, R., Croiset, G., Kruitwagen, C., ten Cate, O.: Validity evidence for the measurement of the strength of motivation for medical school. Adv. Health Sci. Educ. **16**, 183–195 (2011)
19. Koh, C., Tan, H.S., Tan, K.C., Fang, L., Fong, F.M., Kan, D., Lye, S.L., Wee, M.L.: Investigating the effect of 3D simulation based learning on the motivation and performance of engineering students. J. Eng. Educ. **99**, 237–251 (2010)
20. Gero, A.: Engineering students as science teachers: a case study on students' motivation. Int. J. Eng. Pedag. **4**, 55–59 (2014)
21. Lumsdaine, E., Lumsdaine, M., Shelnutt, W.J.: Creative Problem Solving and Engineering Design. McGraw-Hill, New York (1999)
22. Stephan, E.A., Bowman, D.R., Park, W.J., Sill, B.L., Ohland, M.W.: Thinking Like an Engineer: An Active Learning Approach. Prentice Hall, Sadle River (2011)
23. Guay, F., Vallerand, R.J., Blanchard, C.: On the assessment of situational intrinsic and extrinsic motivation: the situational motivation scale (SIMS). Motiv. Emot. **24**, 175–213 (2000)
24. Ryan, R.M., Connell, J.P.: Perceived locus of causality and internalization: examining reasons for acting in two domains. J. Pers. Soc. Psychol. **57**, 749–761 (1989)
25. Deci, E.L., Ryan, R.M., Williams, G.C.: Need satisfaction and the self-regulation of learning. Learn. Individ. Diff. **8**, 165–183 (1996)
26. Schunk, D.H.: Self-efficacy and academic motivation. Educ. Psychol. **26**, 207–231 (1991)

Interrupts Become Features: Using On-Sensor Intelligence for Recognition Tasks

Kristof Van Laerhoven and Philipp M. Scholl

Abstract Wearable sensors have traditionally been designed around a micro controller that periodically reads values from attached sensor chips, before analyzing and forwarding data. As many off-the-shelf sensor chips have become smaller and widespread in consumer appliances, the way they are interfaced has become digital and more potent. This paper investigates the impact of using such chips that are not only smaller and cheaper as their predecessors, but also come with an arsenal of extra processing and detection capabilities, built in the sensor package. A case study with accompanying experiments using two MEMS accelerometers, show that using these capabilities can cause significant reductions in resources for data acquisition, and could even support basic recognition tasks.

1 Introduction

Processing of wearable sensor data has typically been done at two platform levels: one where a micro controller close to the actual sensor does the acquisition, and the back-end processor which typically does more complex preprocessing and classification of the data. Transducers and sensing circuitry embedded in the sensor have mostly been used as a static component that produces values that are regularly queried by the micro controller platform. In recent years, however, many sensing integrated circuits (ICs) have become capable of feats beyond conditioning and digitising the sensor's signal.

K. Van Laerhoven (✉) · P.M. Scholl
Embedded Systems Lab, Department of Engineering,
University of Freiburg, Freiburg, Germany
e-mail: kristof@ese.uni-freiburg.de

P.M. Scholl
e-mail: pscholl@ese.uni-freiburg.de

R. Szewczyk et al. (eds.), *Embedded Engineering Education*,
Advances in Intelligent Systems and Computing 421,
DOI 10.1007/978-3-319-27540-6_12

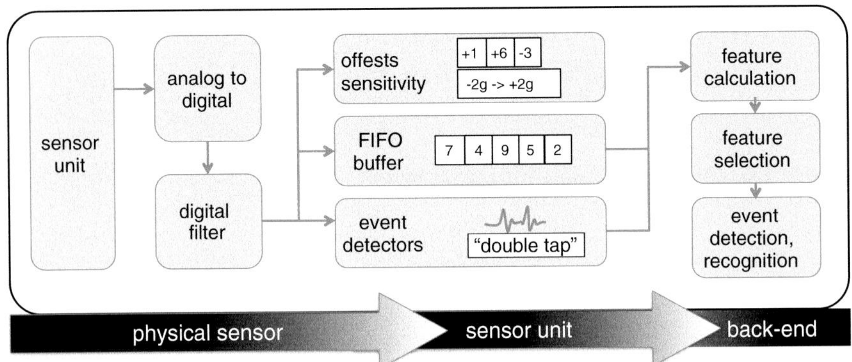

Fig. 1 Calibration, collection, and detection routines (middle three blocks) can increasingly be off-loaded from the unit's micro controller to the actual sensing chip

The outset of this paper is an emerging trend where sensing ICs not only include embedded signal filters, calibration, and conversion routines to digital data, but also contain an increasing amount of built-in detectors that locally abstract the data, which previously had to be implemented on the micro controller or processor. This paper investigates with a case study of two Micro ElectroMechanical System accelerometer chips that provide digitization and detectors on inertial signals, what opportunities present themselves when these digitally-enhanced sensors are fully utilized.

Wearable sensing devices have, more than other types of sensor deployments, to cope with extremely challenging requirements. They should preferably be comfortable long-lasting, and their design thus should balance between being small, light-weight, *and* power-efficient. Given that most sensor units are battery-powered, optimizing the sampling and processing of new data is crucial. The relatively high cost of permanently storing or sending information through a wireless channel has led to several approaches that avoid transmission of redundant data, with techniques being suggested to avoid sampling at excessive rates and to minimize sensor-to-back-end traffic. Local methods have been proposed on the sensing unit, for example hierarchical sensing between low- and high-resolution sensors for same modalities, adaptive sampling according to past sensed values, and compressing of time series [1–3]. Others focused on multiprocessor sensor nodes with staged wakeup, using power-efficient micro controllers to wake up high-end micro processors for processing heavier workloads [4] (Fig. 1).

This paper argues that using sensors with integrated modules that perform tasks that previously were done on an attached micro-controller or processor, could lead to a significant paradigm shift. A case study with two examples of these sensors shows what advantages and impact this has on wearable sensing devices in particular in terms of power consumption and development of recognition algorithms.

2 Background

As a particularly strong example of the type of sensors that have evolved a lot in the past decade, the MEMS accelerometer will be used to illustrate and enumerate several of its features in detail with relation to doing the sensing whilst being more power-efficient, more flexible, and more tailored to the application. As an illustration of how advanced these recent inertial sensors have become, two examples will be used as typical examples of such integrated devices. These features will be linked to applicability in wearable sensing research in the next section.

2.1 Evolution of MEMS Accelerometers

Micro ElectroMechanical Systems (MEMS) implementations of sensors have boomed since the mid-90 s as these versions were smaller, lighter, and cheaper than traditional alternatives, while performing the same functions as larger mechanical systems. Especially accelerometers have gained a tremendous amount of traction as one of the early adopters of MEMS technology, allowing to sense both tilt and motions in a small package. Table 1 shows several accelerometers in the lower-g range, many of which have been used in wearable sensors research,

Table 1 A list of some popular MEMS accelerometers, showing rapid improvements in miniaturization and digitization

Year	Name	Size (mm)	Axes	Output	I (uA)	Detectors	Range (g)
1995	ADXL05	$10 \times 10 \times 4.5$	1	Voltage	8 k–800	-	1–5
1999	ADXL202	$10 \times 7.4 \times 3$	2	Duty cycle/volt.	600	-	2
2003	LIS3L02AQ	$7 \times 7 \times 1.8$	3	Voltage	850	-	2,6
2006	ADXL330	$4 \times 4 \times 1.45$	3	Voltage	320	-	3
2007	SMB380	$3 \times 3 \times 1$	3	SPI, I2C, 1 int.	200	Freefall, motion	2,4,8
2007	LIS331DL	$3 \times 3 \times 1$	3	SPI, I2C, 2 int.	290	Freefall, motion, taps	2,8
2009	ADXL345	$3 \times 5 \times 1$	3	SPI, I2C, 2 int.	145	Freefall, motion, taps	2,4,8,16
2011	BMA250	$2 \times 2 \times 0.95$	3	SPI, I2C, 1 int.	139	Freefall, motion, taps, turn	2,4,8,16

We investigate the usage of an increasing set of on-sensor detectors and their potentials for wearable sensing. Built-in modules detect events such as free-fall (*freefall*), motion change (*motion*), taps or double taps (*taps*), and orientation changes (*turn*)

and produced by a few companies: Analog Devices (prefixed with ADXL), Bosch (prefixed with BMA/SMB), Kionix (KX), and ST-Microelectronics (LIS). As the original applications for accelerometers were mainly targeted toward the automotive industry, in particular airbag deployment, first widely available units were single-axis and required a decent amount of power.

With the introduction of additional markets, such as user interface control in game consoles, smart phones, tablets, and portable computers, a further drive was given to make the devices smaller, more power-efficient, and easier to interface digitally to micro-controllers. The packages of current accelerometers have reached a size small enough to enable integration in the tiniest devices such as wristwatches, lightweight music players, and body sensors. Although the figures for power consumption were extracted from the respective products' data sheets where they are often listed for the lowest supply voltages, these do confirm also a firm drop over the years. All but the very specialized commercial accelerometers have nowadays both Serial Peripheral Interface (SPI) and Inter-Integrated Circuit (I2C) bus interfaces to connect them to a micro-controller, as well as at least one interrupt pin. This also reduces several requirements (no ADC conversion unit, no signal conditioning) for the circuitry around the sensor.

Some clear trends that are visible in Table 1 are the miniaturization and early increase to full 3-axis sensing per package, while a digital interface with programmable interrupts has at present day become the de facto standard. Power consumption has dropped significantly as well, together with a decline in the common supply voltage (from 5 to 1.8 V for the most recent devices). There are several indications that some of these highlighted trends will continue still in the coming years. Several sources [5] have already suggested that 1 mm^2 packages are expected to appear on the market, leading to a further factor-of-four reduction in size. Current MEMS accelerometers use approximately 19 % of the package's area for the actual sensor [5], this would be the bigger hurdle in downsizing the packages even more. From a market point of view, both competition and application demand are still driving toward new and further MEMS accelerometer improvements.

2.2 The New Functionalities

Although several of the latest inertial sensors contain the digital modules discussed here, we have opted to discuss those of accelerometers (such as the ADXL345 and the BMA150) as they both have promising features for wearable recognition tasks.

Calibration. Several recent MEMS accelerometers have an interface to change the sensor's sensitivity via its sensing (g-) range. Just a few years ago, the application dictated the sensor to be chosen at the time of hardware design, in particular the range of gravity they could sense. For the ADXL202, for instance, an absolute acceleration over 2 g would max out the sensor's output, leaving only lower parts of the signal from sudden impacts such as punches, knocks, or hits. Allowing the micro-controller to vary the sensor's gravity range while in operation, postpones this decision from the hardware design phase to the run-time of the system. The sensor can be instructed to dynamically change its sensitivity range, with the trade-off that for higher ranges the values will have lower resolution and that for lower ranges values get capped off. The ADXL345 and BMA150 allow different range profiles: from zero g to ± 2 g, ± 4 g, ± 8 g, and for the former even up to ± 16 g. These overlap with numbers reported in [6].

Apart from an adjustable range, current accelerometers also provide for configurable offsets, allowing calibration of the sensor at run-time whereby the (X, Y, Z) offset values are kept in the sensor. An integrated low-pass filter can furthermore be controlled via the bandwidth parameter, going from 1 kHz down to 8 Hz for the BMA150.

Collection. Another functionality that appeared with the digital interface of MEMS accelerometers is on-sensor temporal storage in the form of an internal First In, First Out (FIFO) buffer for the data samples. This results in a more relaxed interface between sensor and the micro-controller as the latter can spend longer periods in sleep-mode: The ADXL345 has for instance a buffer of 32 samples, meaning that for 10 ms samples, the micro controller needs to accept three larger data packages per second from the sensor, instead of 100 smaller ones. Meanwhile, the micro controller could spend time in a lower-power mode (idle or sleep) or handle other tasks.

Setting up the FIFO is done with two parameters: the first is the option to size the FIFO to n (with $n \leq 32$), the second is the operation mode of the buffer. The latter can be either 'stream mode', in which the last $n \leq 32$ samples are stored in the FIFO while discarding older samples, 'trigger mode', in which only the first $n \leq 32$ samples are stored after a pre-defined event occurs, and 'FIFO mode', in which it stops collecting new samples after the FIFO is filled up. As a default however, the FIFO is set to 'bypass', in which it forwards its samples without logging in the buffer. The buffer option can be combined with a lower-power mode.

Detection. When acceleration data is being captured in a wearable setting, it is very likely that successive sample readings remain unchanged. For most current inertial sensing implementations, the same data are thus repeatedly being communicated, causing redundant samples that only can be avoided by the micro controller running algorithms such as run-length encoding (RLE). Such an approach first of all does not allow the micro controller, or further components in the data chain, to be in idle- or sleep-mode for longer than the sampling interval, and demands more processing to occur at later stages when the data have to be merged. This feature has now emerged in the sensors.

Instead of *detecting activity*, the accelerometers can also be set to *detecting inactivity*, often also per axis. This function can be used together with an auto-sleep function, that configures the accelerometer to go to a low-power sleep mode when no activity is detected, leaving also the micro controller to reside in sleep mode till motion is detected. Parameters for this mode are an activity threshold (if a higher acceleration is sensed, the interrupt is triggered with the activity event),

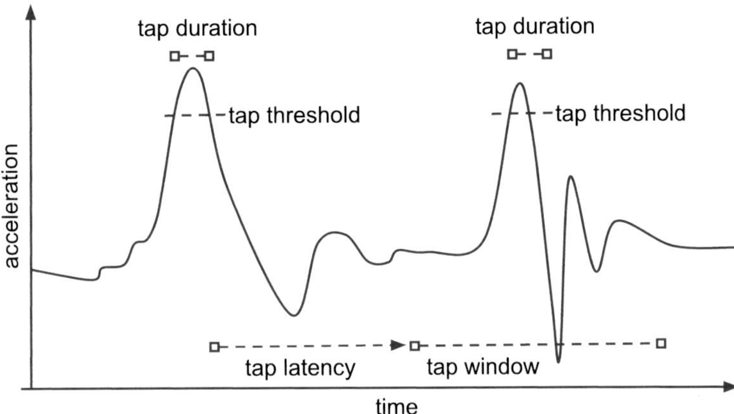

Fig. 2 The parameters that characterize when a tap or double-tap event is detected. After these detectors are set up, interrupts to the micro-controller are sent to signal new detections

and the inactivity time and inactivity threshold (if a lower acceleration is sensed for longer than the inactivity time, the interrupt is triggered with the inactivity event). One of the earlier motivations for including inertial sensing in computing equipment has been to safeguard the local hard disk by detecting whenever the device falls. The *free-fall detection* behavior is defined by a free-fall lower threshold and a free-fall time duration parameter, the former defined for all axes simultaneously and the latter being designed to be typically very short (maximally 1.28 s). The BMA also provides *high-g detection* that triggers an interrupt when the acceleration exceeds a threshold for a long enough period.

As accelerometers have become mainstream components in portable computing devices, their data have also been used for detecting basic interaction. The current dedicated functions on-board new MEMS sensors include the *detection of the user tapping or double-tapping* the device, in order to establish basic input and avoid mechanical buttons. Figure 2 illustrates the parameters needed for tap and double-tap detection.

2.3 Promises for Wearable Sensors

The evolution of MEMS accelerometers went through profound changes in the last decade. We listed so far several of the typical advanced features of the most recent units. As with many popular components, these sensors have dropped in price significantly, and their component size and power demands have similarly, incessantly been reduced as newer types were produced. We argued that on top of these improvements, a set of new capabilities have emerged by the introduction of [7] *allowing run-time calibration and sensitivity re-configuration*, [5] *sensor-internal buffering*, and [8] *sensor-specific detection features* that would be costly when implemented on the unit's micro controller.

Calibration. Having the option of setting the axes' offsets, low-pass filter bandwidth, and the sensitivity range of the accelerometer from within an embedded system, allows detection processes to direct the system's sensitivity as they see fit. To cite an extreme use case, [9] for instance observed that extreme accelerations could happen for wearable sensors used during baseball pitching, while other interesting activities might rely on slight muscle tremors in the *mg* range. Accelerometers that detect both would be expensive and produce high-range values that would be harder to analyze.

Collection. The use of a sensor-integrated buffer has consequences on two properties of any system subsuming the sensor: Processing ticks are taken from the host and moved to the sensor so that the micro-controller can take on additional duties. The reduced sensor-host communication can lead to power saving opportunities when the micro-controller uses the time between communication to sleep. The buffer feature impacts several feature calculation choices that otherwise would be implemented on a buffer in the micro controller's RAM: Some of the most popular descriptive features to abstract sensor data are basic statistics that operate on a buffer, such as mean and variance [8]. When the buffer is sensor-managed, one-pass implementations that calculate mean and variance on the fly might be used to avoid having to buffer the sensor data on the micro controller at all. Assumed that sensor's buffer length is sufficient (the 32 samples stored in the ADXL345 might for example not be enough yet), benefits to the micro-controller would be a further reduction in processing and memory requirements.

Detection. By using the activity and inactivity detection modules in the inertial sensor, thresholded run length encoding can be implemented on the sensor rather than on the micro controller. For many wearable applications that rely on detection of infrequently occurring events that are detected by sudden changes in the sensors' signals, the nodes' micro controllers can be placed in sleep mode for most of the time, only to be woken up when these events occur. The threshold parameter could be used to either cope with noise in the sensor signal to send an interrupt only when real changes occur, or as an actual rule to detect events that exhibit a signal above given fluctuations. The tap and double tap detection facilities of the sensor could be particularly helpful in the analysis of more complex signals that contain characteristic peaks and harmonic signals. The tap duration and tap threshold parameters effectively instruct the accelerometer to respond only to an impulse over a given amplitude and length, the double tap detection feature could similarly be used to detect two such following occurrences.

This paper contains two studies: A first study concentrates on improvements achieved for the collection of wearable inertial data, using the sensor's buffer and implementing RLE on the sensor chip. A second study focuses on supporting wearable recognition.

3 Experiment Setup

As a platform for testing the impact of individual features introduced in the previous section, we use a custom-built platform around a Microchip 18F46J50 micro controller, which includes all necessary components in a small-scale form-factor to sense and analyze inertial data. Connectors and circuitry are available for storage of data on a local μSD card, as well as an OLED display for visualization and providing more advanced debug information. This unit was designed specifically to capture wearable inertial data at a relatively high frequency for activity recognition purposes, and its micro controller contains several advanced modules for USB communication and interfacing with persistent storage (up to 2 GB on an attached μSD card, made accessible as an external disk via the USB connector). The prototype with battery is depicted in Fig. 3.

For the current draw analysis, it is important to note that the micro controller is able to switch between an external 12 MHz high-speed crystal (internally quadrupled to 48 MHz using Phase Lock Loop) and an internal clock circuit that can be set between 31 kHz and 8 MHz. There are 64 KB available in program memory and 3776 bytes for RAM. The battery is connected to a charger and voltage regulator circuit that supplies the node with a constant 3.3 V, whether powered from USB or battery. Although sleep modes are available that drop the current demand for the micro controller alone to 15 nA, applications depend on an internal real-time clock maintaining its state, requiring 830 nA. Furthermore, since an interrupt-triggered wake-up time from deep sleep is rather slow, we used the standard sleep mode. To obtain an as realistic as possible measurement for a wearable setup, the prototypes were powered from miniature Li-Polymer rechargeable batteries with a capacity of 180 mAh that were fully charged, from the prototype's charging circuit, before the experiments started. The current draw was monitored by a digital multimeter with sub-milliamp measuring facilities between battery and main

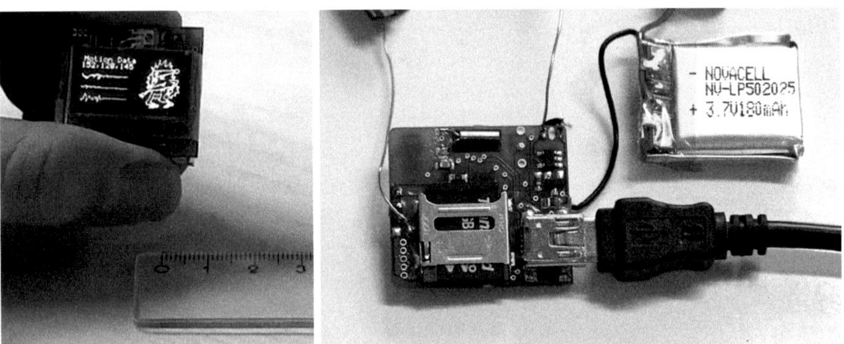

Fig. 3 The sensor prototype (*left*) is fitted with a 3D inertial sensor, OLED display, and SD card. Right is the close-up of the prototype in the experiment setup, connected to a miniature battery for current draw and timing measurements. Components irrelevant to the experiment were disabled

board. For timings, a set of test points and free digital output pins on the prototype unit were connected to a digital oscilloscope to monitor timings of communication between micro controller and its connected components.

Although taking actual current draw and timing measurements deliver the more realistic figures when compared to simulation or numbers provided in data sheets, the downside is that for this prototype the results will be anecdotal and restricted to one particular platform. However, the presented comparisons and conclusions should be still indicative for many wearable sensors, since the platform and its micro controller follow the traditional design of others (for instance, the micro controller has similar features and properties to those in [10], or the same sensor as in [7]).

4 Evaluations

In this paper, two questions are asked: (1) what benefits do the new built-in digital functions of the discussed sensor chips offer for data sampling? and (2) how can on-chip detectors be used in recognition tasks?

4.1 Sensor Unit Configurations

The first experiment's aim is to observe the impact of the sensor-based FIFO sampling and the motion detection, instead of micro controller-driven sampling with run-length encoding. The aim of the second experiment is a comparison in recognition performance between peak features calculated on the micro controller and the sensor-based tap detectors. All setups were using the prototype with the ADXL345 and BMA150, and configured in a raw data sampling scenario and a feature preprocessing scenario.

Baseline sampling In the traditional sampling setup, the micro controller queries the ADXL345 and BMA150 every 10 ms[1] via SPI for a new data, stores these new data with run-length encoding, and goes into sleep mode for the rest of the time to be woken up by an internal timer. This includes occasionally storing full buffers on μSD together with time stamps for later synchronization, hence the need for the real time clock to stay operational.

FIFO sampling. Since only the ADXL345 contains this feature, a second configuration uses this sensor only, in the FIFO buffer 'trigger' mode with the maximum size (32 samples). All further parameters are set as in the baseline setup, i.e., the sampling is set to 10 ms and after the entire FIFO is sent to the

[1]Leading to 100 Hz sampling rate (as also used in e.g. [11]).

micro-controller, all data is stored via run-length encoding and with time stamps on the μSD card.

On-chip RLE sampling. A third setup uses the motion detection of both accelerometers with the same sampling parameters as in the other two setups, i.e., the sensor data is also in this setup recorded with time stamps on the μSD card, with the run-length encoding (RLE) performed on the sensors instead of executed on the micro-controller.

Micro-controller-feature recognition. The first recognition setting is implemented via features that are calculated on the micro controller and stored on the μSD card for later classification. For peak detection, an algorithm similar to the one suggested by [?] was used with the best parameter settings after extensive evaluation.

Tap-detector recognition. As an alternative, a second setup uses the ADXL345's tap detection as a feature. The tap parameters were set to be sensitive enough to detect significant peaks in the signal (duration: 72, threshold: 90), and taps, along with axis in which the tap was detected, were stored by timing the intervals in 10 ms increments.

4.2 Sampling Evaluations

FIFO sampling. The differences between regularly sampling the accelerometer every 10 ms from the micro controller, and waiting for the ADXL345's buffer to be filled (signaled by a so-called *watermark* interrupt), are twofold: [7] the micro controller is able to switch to a sleep mode for a longer time, and [5] the transmission of the buffer will be done in larger bursts instead of single packets (approx. 3 times per second in this case). The evaluation shows that the time taken to read the entire buffer via SPI is at most 3.1 ms, while the processing of the buffer takes an additional 2.1 ms. As these still comfortably fit in the 10 ms window between readings, even larger buffers than the currently available 32 three-dimensional samples would be possible at 100 Hz. Two observations stand out in Fig. 4: the use

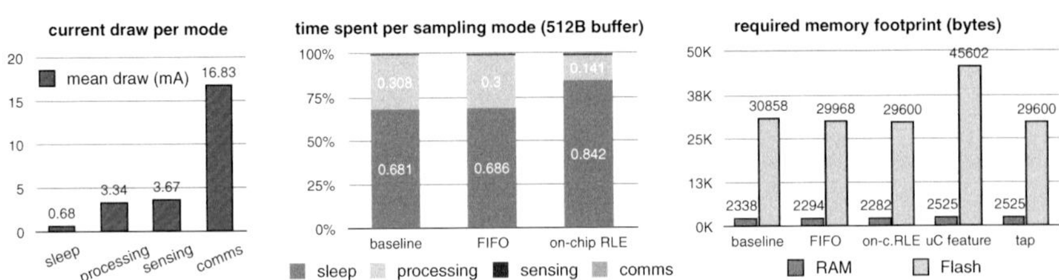

Fig. 4 *Left* The node's power draw measured for each mode (low-power sleep, micro controller processing, sensor data acquisition, and memory buffer logging). *Middle* amount of time spent in each mode for sampling to a 512-byte data buffer, for each of the three sampling setups. Right: The node's firmware footprint in RAM and flash memory required, for each of the 5 setups

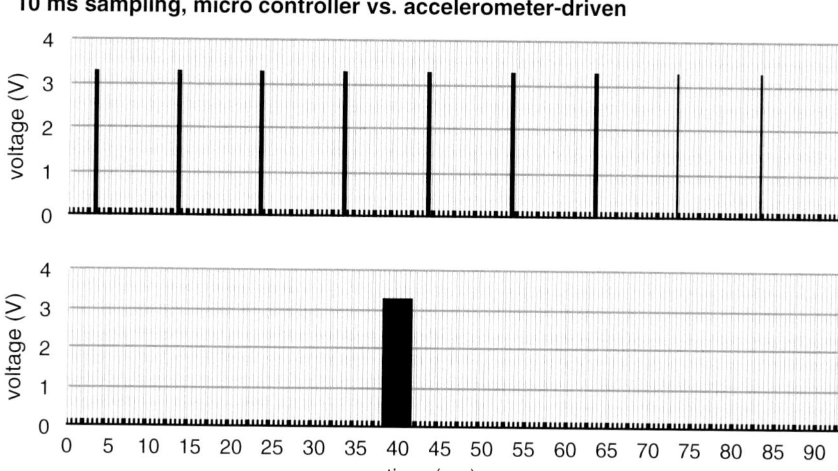

Fig. 5 Timings for micro controller-driven sampling (*top*) and sampling on the ADXL345's FIFO buffer (*bottom*) as observed by an oscilloscope. High 3.3 V states indicate SPI communication

of the FIFO leads to only a slight performance increase in terms of communication (reduction of 12 % in SPI traffic) and amount of code (a 2–3 % smaller size). The overall power consumption however is nonetheless reduced by 4 %. Figure 5 illustrates the interaction between micro controller and accelerometer: code markers were used for SPI timing measurements, which pulled up digital monitor pins of the micro controller, and which were timed by the oscilloscope. The micro controller does not need to query for a new sample each 10 ms, but instead accepts the entire buffer every 320 ms after waking up from an interrupt. There is only a slight reduction in program memory and RAM requirements when using the ADXL345 internal buffer, since storage is still required on the micro controller for buffering to the μSD card, leading to only 2 % less RAM usage in comparison to the baseline approach and a 3 % reduction in program code size. For the power-usage of the prototype, the mean-measured current draw (taken continuously over 100 ms) is more significant: it dropped from 0.690 to 0.660 mA. As the timing is not that different, this drop could be due to the ability of the ADXL345 to save power, as it handles less SPI requests.

On-chip RLE. The difference between the baseline approach and using the on-chip change detection has similar advantages for both sensors, but especially power consumption dropped considerably (18 %). The former's readings are visualized in Fig. 4: A slight drop can be seen in the memory footprint for the on-chip RLE setup (4 %), as the run-length routine and temporary variables are not required anymore. More time was spent in sleep mode: Since this setup was confined to a lab bench, this might be an optimistic value (as the prototype was not heavily moved). Nevertheless, the current draw going from 0.690 to 0.580 mA during low activity is significant.

4.3 Recognition Evaluations

For evaluating the performance of the tap features for classification, and comparing them with similar features implemented on the micro-controller, two prototypes were worn next to each other on the dominant wrist of a participant, for a 24 h period. The two prototypes were synchronized: one was sampling raw data together with the tap features, the other was set up for collecting the micro-controller peak features. Both features abstract accelerometer data by the amplitudes of larger peaks in the signal, the widths of these peaks, and the regularity at which they occur. The tap detector provided by the ADXL345 detects only one particular type of configuration for the tap duration, threshold, latency and window, so a lower bound was chosen, and solely counted the amount of interrupts generated by the accelerometer over a window of 3 s, and characterized by the axis which caused the interrupt.

The data from five performed target activities, listed in Table 2, was separated from the entire data set with approximately 2 min of transition data immediately before and after each activity removed, and each activity's data split in 5 folds for leave-one-out cross-validation against the remaining whole data set including the background data. Classification was performed using the Support Vector Machine (SVM) classifier. Figure 6 shows the five target classes embedded in the full tri-axial acceleration data set.

The right plot in Fig. 4 shows the total memory footprints for the setup where the feature code is implemented on the micro controller and when configuring the ADXL345 accelerometer to send interrupts for every detected tap. The adding of the feature detection routines on the micro controller particularly require a significant amount of program memory. The collecting of tap features performs slightly

Table 2 The list of target classes to be detected, along with their precision (pr.) and recall (rec.) for micro-controller-based features (uC) and using the accelerometer-based tap detection (tap)

Pattern	uC	uC	Tap	Tap
To detect	pr.(%)	rec.(%)	pr.(%)	rec.(%)
Nordic walking	68.09	52.22	35.20	54.45
Running	74.99	95.42	71.55	92.87
Bicycling	71.82	77.84	52.85	75.68
Typing	60.20	85.78	68.83	84.44
Badminton	85.37	97.67	70.00	97.24

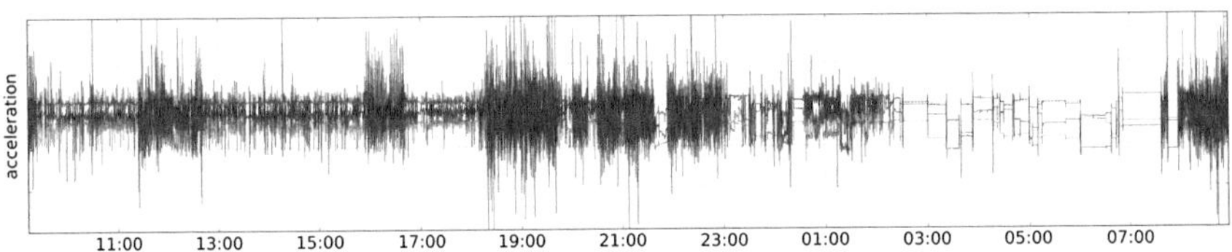

Fig. 6 A 24 h data set for evaluation the tap feature, while the prototype was worn on the wrist. 5 activities were performed for about an hour each, among 20 h of background data

worse than the on-chip RLE setup in terms of timing and power efficiency, but also in this case there might be a bias as these values were measured in the lab bench setup. The precision-recall values in Table 2 show that for most activities, the features on the micro controller performed similar to using the tap feature on the accelerometer together with the times between interrupts and the axis on which they were generated. Overall the on-micro-controller features perform better, for "bicycling" and "nordic walking" even much better, since it has several advantages still, one of them the 3 s window over which peaks are evaluated being larger than the ADXL345 allows for its tap detection.

4.4 Summary of Results

This evaluation section has investigated the benefits of using the built-in features of a recent inertial sensor for one specific, yet typical, sensor node. Measurements included [7] timing information on the SPI communication between sensor and micro controller, [5] the code size required to implement them, [8] detection precision and recall values, and [9] the overall current draw of the entire prototype sensor node. The results gathered in this evaluation can be summarized in two parts. For the off-loading of the buffering and sampling to the MEMS chip, the conclusions of this case study are that it leads to:

- more efficient communication between sensor and micro controller, as witnessed in the communication (reduced to 3 sending-bursts lasting 3.1 ms each)
- less buffer variables and routines to be implemented on the micro controller, and a slightly smaller code footprint and reduced RAM requirements (2 %, resp. 3 %)
- an overall drop in the current draw of the entire sensor node (0.58 mA instead of 0.69 mA, mean average draw), making it last longer on a battery charge.

For using the tap detector in the recognition of basic activities in the acceleration signal, a preliminary classification evaluation showed that:

- similar reductions as above are observed, as SPI communication is minimized and peak detection on the micro controller requirings a larger amount of resources
- using tap detection as a feature with minimal resources on the micro controller performs for the tested basic classes only slightly worse. Identified limitations are that thresholds need to be set beforehand and that width of peaks are more limited.

5 Conclusions and Outlook

The future is promising for power-efficient wearable inertial sensing: new generations of MEMS sensors have started to move several basic detection techniques and added functionality in their sensors that can immediately be used for the detection of patterns and features in the sensed signal. This paper showed by means of a case study involving wearable activity sensing that the benefits lie in a lower threshold for implementing detectors on the node's micro controller, and a more power efficient design overall.

Three promising areas were identified in a case study using the ADXL345 and BMA150, showing that the benefits are worth taking into consideration for many wearable applications. The use of a sensor-internal buffer reduces resources on the micro controller, setting the sensitivity of the sensor (g range and bandwidth in this case) enables new applications at a slight cost of communication overhead, and on-sensor detection routines allow processing to happen before the micro controller receives the data. This leaves less constraints on the micro controller, so that a lower-power version can be chosen (in the spirit of [12]), or the extra resources can be utilized to do advanced data or communication processing, or simply spend more time in sleep mode.

New products have been announced recently merge acceleration and other modalities, like 3-axis gyroscopes and magnetometers, including the recognition of basic activities like sleep or walking. Although the size and energy requirements for these are still an order of magnitude larger than the MEMS devices discussed in this paper, it is foreseeable that soon full IMU devices will provide an even richer set of basic detection features, that could equally have an enormous impact on wearable sensing units that can sense and analyze complex motion patterns.

References

1. Rahimi, M., Hansen, M., Kaiser, W., Sukhatme, G., Estrin, D.: Adaptive sampling for environmental field estimation using robotic sensors. In: IEEE/RSJ International Conference on Intelligent Robots and Systems, 2005. (IROS 2005), pp. 3692–3698 (2005)
2. Concha, O.P., Xu, R.Y.D., Piccardi, M.: Compressive sensing of time series for human action recognition. In: International Conference on Digital Image Computing: Techniques and Applications (DICTA), pp. 454–461 (2010)
3. Van Laerhoven, K., Berlin E., Schiele, B.: Enabling efficient time series analysis for wearable activity data. In: International Conference on Machine Learning and Applications. Institute of Electrical & Electronics Engineers (IEEE) (2009)
4. Raghunathan, V., Ganeriwal, S., Srivastava, M.: Emerging techniques for long lived wireless sensor networks. Commun. Mag. IEEE **44**(4), 108–114 (2006)
5. Dixon-Warren, S.: The evolution of compact three axis accelerometers. MEMS Industry Group, MEMSblog (2010)
6. Minnen, D., Starner, T., Essa, M., Isbell, C.: Discovering characteristic actions from on-body sensor data. In: 10th IEEE International Symposium on Wearable Computers, pp. 11–18. IEEE (2006)
7. Zolertia z1 datasheet (2011)
8. Huynh, T., Schiele, B.: Analyzing features for activity recognition. In: Proceedings of the 2005 Joint Conference on Smart Objects and Ambient Intelligence, pp. 159–163
9. Lapinski, M., Berkson, E., Gill, T., Reinold, M., Paradiso, J.A.: A distributed wearable, wireless sensor system for evaluating professional baseball pitchers and batters. Wearable Computers, IEEE International Symposium 0, 131–138 (2009)

10. Polastre, J., Szewczyk, R., Culler, D.: Telos: enabling ultra-low power wireless research. In: Proceedings of the 4th International Symposium on Information processing in Sensor Networks, pp. 48–es. IEEE Press (2005)
11. Lorincz, K., Chen, B.R., Waterman, J., Werner-Allen, G., Welsh, M.: Resource aware programming in the pixie os. In: Proceedings of the 6th ACM conference on Embedded network sensor systems. pp. 211–224. SenSys '08, ACM, New York, NY, USA (2008). http://doi.acm.org/10.1145/1460412.1460434
12. Roedig, U., Rutlidge, S., Brown, J., Scott, A.: Towards multiprocessor sensor nodes (2010)

Printed in the United States
By Bookmasters